Sensory Exotica

Sensory Exotica
A World beyond Human Experience

Howard C. Hughes

A Bradford Book
The MIT Press
Cambridge, Massachusetts
London, England

First MIT Press paperback edition, 2001

This book was set in Sabon by Wellington Graphics and was printed and bound in the United States of America.

Library of Congress Cataloging-in-Publication Data

Hughes, Howard C.
Sensory exotica: a world beyond human experience / Howard C. Hughes.
p. cm.
Includes bibliographical references and index.
ISBN 0-262-08279-9 (hc.: alk. paper), 0-262-58204-X (pb)
1. Senses and sensation. 2. Physiology, Comparative.
3. Echolocation (Physiology) 4. Electroreceptors.
5. Magnetoreception. 6. Pheromones. I. Title.
QP435.H84 1999
573.8′2—dc21 98-51875
CIP

Illustration credits are found on p. 317.

This work is dedicated to the two women in my life. To my mother, Jeanette, who for my first 21 years tried (successfully, I think) to teach me the important things in life. And to my wife Katherine, who for the last 22 years, has made sure I don't forget what those important things are.

Contents

Preface

There are a couple of unique things about this book. Well, at least they are unique for me. For one thing, this is my first book project. While a reasonable reader might legitimately not care about that, I mention it in the hope of obtaining some allowances for "rookie mistakes" that may have escaped my notice—or indeed, my ability to detect rookie mistakes in the first place. Second, the work described in these pages was done entirely by others. Although I have aspired to be a scientist all of my adult life, and have been fortunate enough to make a living doing science and teaching others about science and the scientific approach to understanding the natural world, I had nothing to do with the remarkable insights I am attempting to relate. I feel somewhat like a network news anchor—I didn't make these stories of discovery, I'm only trying to tell them. I feel compelled to say that partly out of the sense of admiration I have for the many gifted and dedicated people who did make these discoveries. I thank them for their gift—a gift of understanding and insight. And I hope that this book is at least a small tribute to their legacy of intelligence, creativity, perserverance, and sheer hard work.

As I said, I feel a little like a journalist who is telling a story that is not of his own making. The story is intended to be accurate, but it is also intended to be entertaining. That can be a difficult balancing act, but that is what I've tried to do. The book itself grew out of a course I teach at Dartmouth College. For 18 years I have taught undergraduate and graduate students about the workings of our own magnificent senses—vision, hearing, touch, taste, and smell. I have been fascinated by the mechanisms of eyes and ears—and of the neural systems that support our perceptions—since I was a college student. I feel fortunate that I've been able to pursue these interests while remaining gainfully employed. It's almost like having someone pay you for indulging in a hobby. Several years ago, I thought it might be fun to add a little twist to the curriculum by offering a course that dealt with senses that animals have but that, according to most, people do not have. That course eventually led to the writing of this book.

Teaching at Dartmouth is an exciting and challenging enterprise. The students are very talented, and many have a deep-seated fascination with the nature of things that is so pervasive in their lives that it is an essential characteristic of their personality. It is an important part of who and what they are. I have been fortunate to have had many such students in my classes. Teaching them is always a challange. They are active processors of information. But they do not simply learn what you teach them. They evaluate it and explore its implications. Teaching students like that is not for the faint of heart. You have to be willing to face your own limitations—sometimes right there in the classroom, in front of everyone. Sometimes you find you didn't understand something as well as you thought. Sometimes the roles of student and teacher get reversed. Usually, I like it when that happens—so long as it's not a habitual turning of the tables! So I would also like to thank the many Dartmouth students who did library research and often wrote remarkably well-crafted term papers on some of the topics included here. They were an enormous help, and are a continuing source of ideas and inspiration.

Finally, I wish to thank all the talented people at the MIT Press. I have been learning that every author needs skilled editors, and I am no exception. I would like to thank Michael Rutter, the neuroscience editor, for his confidence in me, his enthusiasm for this project, his vision, and his patience. Production editor Deborah Cantor-Adams and designer Chryseis Fox were invaluable to me, and I sincerely thank them each for their efforts. Many thanks are due also to Beth Wilson, the freelance copyeditor who initially edited the manuscript. Finally I wish to thank Beth Adams, my dear friend of many talents, for agreeing to do the artwork on the jacket.

While I have many people to thank, none are responsible for the actual narrative that follows. For the actual telling of the tale, I alone must bear the responsibility. It was an interesting and enjoyable experience writing this work. If you've ever watched the bees visiting your garden, or a formation of geese flying south for the winter, or bats swooping through the skies at dusk, this might give you a little added perspective. And, I hope, a little added enjoyment. If you haven't done any of these things, maybe now you will.

Sensory Exotica

1

Prologue: Perceptions, Misperceptions, and Egocentrism

As a child, I read an old parable whose message has periodically come to mind ever since. I frequently have occasion to tell it to students in classes, and find I need an occasional reminder myself. It's a story about four blind Hindus who had never seen an elephant. One day, they went to visit an elephant for the first time. The first stepped toward the animal, and happened to touch its ear. "The elephant is much like a fan," he concluded. The second touched the leg and thought the first a fool. "Elephants are nothing like fan," he said. "They are more like a tree trunk." The third felt the animal's side, and concluded that elephants are like walls. The fourth touched the trunk, and decided elephants are like large snakes. The message, of course, is that we should not mistake the way things seem to us for the way they really are. We all need to be reminded that our limited perspective can lead to gross misperceptions of the nature of things. Despite the wisdom acquired through age and education, most of us are quite capable of mistaking partial information for conclusive evidence. Few of us can run a marathon, but we are all plenty fit enough to jump to an erroneous conclusion!

People can be paradoxical. We seem to have a real need to explain the phenomena we see. But often we seem curiously insensitive to the fundamental difference between a description and an explanation of natural events. If a child sees the aurora borealis, she might naturally ask, "What's that?" The parent might respond, "Oh, that's the aurora borealis." Many children will be unsatisfied by that answer. Knowing a name is not an explanation, the exceptional child replies, "What is the aurora borealis?" The parent now mumbles something about how the atmosphere creates these nighttime lights, and that a similar effect occurs near the South Pole. Many children (and adults) would find that "explanation" satisfactory.

Notice, however, that nothing has really been explained. The phenomenon has just been described in more elaborate terms. A real explanation would include why the lights are seen only near the poles and not the equator, and how they relate to sunspots. A real explanation must provide an understanding of electromagnetic radiation, Earth's magnetic field, and how these two interact to ionize gas molecules in the upper atmosphere. This in turn requires an understanding of exactly what "ionization of gases" means, and how ionized gases can produce light. The real explanation is pretty elaborate, and it takes a little more effort to grasp. But it does provide us with a much deeper understanding. That deeper understanding is often accompanied by a much more profound appreciation of the nature and beauty of the aurora. I hope this book serves that purpose for a small subgroup of nature's mysteries: sensory modalities beyond the realm of ordinary human experience.

So our concern is with the perceptual world of other creatures, and our understanding is made more difficult because we have no direct way to measure such things. It is hard enough for humans to try to understand the mental life of other humans. We spend a great deal of effort trying to communicate with one another, often with very limited success. It's difficult enough to formulate your ideas or feelings, and try to communicate them accurately to another. A listener might *think* he understands, and may assure you that he really does. But how do you know? How can you be sure? Our mental lives are private, and it's hard to make them public, even when we want to. We do have one ability that contributes immeasurably to our attempts at communication. That, of course is language. Dr. Doolittle not withstanding, that advantage is not available in our attempt to understand other animals, and as a result, misperceptions of even our closest relatives on the tree of life are common.

So if you wanted to know whether birds can sense when a storm is approaching, or whether cats can "feel" that an earthquake is imminent, how would you find out? Clearly, you would have to make such determi-

Figure 1.1
Clever Hans receiving a mathematics lesson.

nations by some aspect of their behavior, and that can be a very tricky business.

Consider, for instance, the case of Clever Hans. Hans was a horse who lived in the late nineteenth century. Reportedly, Hans could count. He not only could indicate his age, or the age of his owner, but he also could perform simple arithmetic operations like addition and subtraction. Billed as the horse that knew simple mathematics, Hans become quite a celebrity. He and his owner toured extensively in Europe, demonstrating these remarkable equine computational abilities to appreciative audiences—for a fee. When asked to perform a certain calculation, Hans would tap out the answer with a front hoof (figure 1.1).

A psychologist named Oskar Pfungst eventually was able to provide some insight as to how this done. Although Pfungst established that anyone could ask Hans the question, he discovered that Hans could provide the right answer only if he could *see* his owner. Apparently, the owner provided subtle visual cues: Hans would simply continue to tap until cued to stop by his owner. A cheap trick perpetrated by a common charlatan? The work of a con artist? Well, the situation was a little more complex and interesting than that. It appears that the owner was completely unaware that he was providing Hans with any cue. He did not *know* exactly *what* he did that told Hans the answer. Whatever cue Hans had learned, it was so subtle that determined human observers could not figure out exactly what it was.

Clever Bertrand provides an even more dramatic example. Bertrand was a contemporary of Hans who also could perform arithmetic calculations (apparently, horses were smarter back then). But unlike Hans, Bertrand was blind! Most experts agreed that the horse must have used subtle auditory cues, although once again the exact nature of these cues were never established (Grier and Burk, 1996).

Other entertaining tales of misperception and misinterpretation of animal behavior and animal intelligence abound. In an article titled "Myths of Animal Psychology," C. O. Whitman adds some interesting cases to what is quite a lengthy list. He describes a vicious attack by a large boar on a 19-year-old boy outside of Rochester, New York. The young man, a Mr. George Howard, was grooming his favorite steed when a large hog broke out of its pen. The boar fiercely attacked young Mr. Howard, who fell to the ground near the horse's feet. Just as the hog was about to inflict the death blow, the horse kicked the hog away. Surely the horse had saved the boy's life!

The story appeared in the *Rochester Union and Advertiser*, and, consistent with the highest standards of journalism, was corroborated by an eyewitness. The headline read, "A Horse Protects His Master from the Tusks of a Savage Boar." Whitman points out, however, that the actual circumstance do not justify the conclusion that the horse was trying to save its master. Rather, Whitman suggests that it was a simple act of self-defense on the part of the horse that was misinterpreted by both the victim and the witness. Whitman's article was printed in 1899.

Now you might think that much has changed in the last 100 years. We are more sophisticated than people were way back then. But in many ways, little has changed. Many people are still quite willing to resort to the mystical or the supernatural in search of their explanations. Take, for example, an article in a Sunday edition of the *San Francisco Chronicle* (April 1996), which describes the curious (and perhaps dismaying) resurgence of dowsing in America. Dowsing is that method of finding water, oil, or gold using a rod. As the dowser approaches an underground spring or vein of gold, the dowsing rod begins to point downward—apparently on its own. Something like a Ouija board. One part of the article describes a man who uses a dowsing rod to communicate with whales. He stands on the high cliffs along the California coastline, and claims to communicate with unseen whales over remarkable distances. Well, you might think this exceedingly unlikely, even absurd. And I agree. By what physical means could such communications possibly take place? Still, many people wish to adhere to beliefs that appear to defy the laws of physics.

There is something about our nature that longs to believe in magic, in things that appear supernatural. But what if I told you that it is very likely that whales can probably communicate with *other whales* over miles of open ocean? Or that many animals have an internal compass that they use as a navigational tool that permits them to migrate over long distances? Or that some fish communicate using coded messages that are sent through the water via electrical fields, and that bees and other insects see things that are completely invisible to our eyes? Sometimes the facts are stranger than any fiction, and, as I hope to show, more remarkable than any magic trick or illusion—because they are true, and because their explanations rely on natural laws.

We know a great deal about how many of these exotic senses work. We understand them in mechanistic terms. Such a mechanistic understanding doesn't diminish the sense of wonder we can have about these elegant systems. Rather, it enhances it.

When we consider sensory perception, we naturally focus on the five "special" senses: vision, hearing, touch, taste, and smell. It is through these senses that we experience the world outside our bodies. There is also the world inside our bodies, and there are sensory organs that provide information crucial to internal body states. Our senses of balance, of body motion, and of posture, depend on sensory organs in the inner ear, in our joints, and in muscles. There are even organs that monitor such things as the levels of carbon dioxide in the blood, blood pressure, and blood glucose levels. These organs provide the brain with information essential to life, but they do not produce conscious sensory experiences (otherwise, people would be aware of the onset of hypertension, and it would less frequently go untreated).

Rather than ESP, perhaps we should call these internal sensory systems our sixth sense—a sense beyond the more familiar modalities of vision, hearing, touch, taste, and smell. If so, then this book is about the seventh, eighth, ninth and tenth senses. What are these new sensory modalities? Well, first of all, they are not new. Their possessors have been relying on them for millions of years. It's just that we've "discovered" them only since the 1970s and 1980s. But newness aside, they include such hi-tech systems as biological sonar systems, sophisticated navigational systems, and senses based on electrical fields.

These systems initially seemed so unlikely, so incredible, that many were reluctant to believe they existed. But they do. And in just a couple of decades we have learned a great deal about the detailed working of these remarkable sensory systems. In some respects, the mechanisms are not very different from the more familiar mechanisms of vision, hearing, or touch. In other ways, the differences are quite dramatic. They have one thing in common, however: they all provide vivid illustrations of the creative genius inherent in the process of evolution.

When we speak of sensory experiences beyond the realm of our five special senses, many may think of supernatural things like extrasensory perception: clairvoyance (the ability to see that which is not visible) or telepathy (the ability to sense the "thought waves" of others). Some have suggested that there may be a resurgence of mysticism in modern society, which is curious because there is not a single case of extrasensory perception that has withstood the tests of rational analysis and rigorous experimental control. But, as we shall soon see, the workings of biosonar, electroreception, and other exotic senses are far more interesting than bending spoons and reading unseen numbers on a stranger's driver's license.

Imagine a sonar system more sophisticated than that found in our most advanced submarines. Now imagine that system is used by a small bat that easily fits in the palm of your hand. All the computations that permit the bat to identify the distance, the speed, and even the particular species of insect target are performed by a brain that is smaller than your thumbnail! That is a truly remarkable device. But it is a device. It can be understood in mechanistic terms. Despite all the folklore associated with bats, no appeal to forces beyond natural laws is necessary. And that is what is truly remarkable about these most interesting creatures.

Each of our senses is a wondrous system of information processing. The events that culminate in perception begin with specialized receptor cells that convert a particular form of physical energy into bioelectric currents. Different receptors are sensitive to different types of energy, so the properties of the receptor cells determine the modality of a sensory system. Ionic currents are the currency of neural information processing, and current flows that begin in the receptors are transmitted through complex networks of interconnected neurons and, in the end, result in a pattern of brain activity we call perception. We can distinguish a red 1957 Chevy from a blue 1956 Ford because each car produces a different pattern of neural activity.

The percepts that result from all this brain activity usually provide us with an astonishingly accurate window through which we view the outside world. If that were not so, we couldn't hit a curve ball, or teach kids how to catch one (you've probably noticed they invariably need instruction). In short, our interactions with the world would not be possible. We interact with our environments so effectively and so effortlessly, it is difficult to appreciate the extensive computations that underlie even the simplest sensory experience. We become convinced that we see "what is really out there." But there are some refractive errors in our sensory windows to the world, some distortions. And our perceptual windows are not as transparent as we think.

For instance, our sense of vision depends upon wavelengths of light that range from about 430 to 700 billionths of a meter. But the entire electromagnetic spectrum covers a range that is approximately 300,000,000,000,000,000,000,000,000,000,000 times larger than what we call the visible spectrum! Clearly, there is a (very large) portion of the electromagnetic spectrum that we cannot detect. That doesn't mean the energy isn't there, and it doesn't mean that other creatures cannot detect portions of it that to us are not visible. Our visual receptor cells are sensitive to an incredibly small range of wavelengths. Other animals are endowed with different types of receptors that render them sensitive to portions of the spectrum that we cannot see. Those wavelengths that are a little shorter than what we call blue light are called ultraviolet wavelengths, or UV light. We can't see UV light, but some insects can, and they use their UV sensitivity as an aid to navigation. In contrast, wavelengths that are a little longer than what we call red light are called the infrared wavelengths. Infrared (IR) radiation is emitted by warm objects. Have you seen the special night goggles that allow people to see in the dark? They work because they have detectors that are sensitive to the infrared part of the spectrum. Any object that is warmer than its surroundings will produce an IR "signature," and the goggles convert those infrared emissions to wavelengths of light that we are able to detect.

Certain snakes evolved their own form of IR night goggles: little pits that act like pinhole cameras for infrared radiation. Snakes use their infrared system to detect and localize their warm-blooded prey. The system was discovered when it was recognized that an agitated rattlesnake will produce accurate strikes at a warm soldering iron, even if its eyes are covered. All members of the family of venomous snakes known as pit vipers have these infrared detectors—it is the infrared-sensitive pits that give them their name. As far as we know, only two types of snakes have this infrared sensitivity—the pit vipers and some boid snakes (constrictors like the boa constrictor). It would probably be of no use to warm-blooded creatures. Their own body heat would produce so much noise in the system that detection of other objects would not be possible. For this system to work, the animal has to be a cold-blooded hunter . . . as cold as the desert night. Rattlesnakes have been tracking their prey by the heat trail they leave behind for millions of years.

Analogous differences in the range of hearing exist for different animals. Dog whistles are one familiar example: we can't hear the whistle, but dogs can. The reason is that they are sensitive to a higher range of sound frequencies than humans are. Yet the auditory abilities of dogs pales in comparison with that of bats or dolphins. Some bats can actually hear the footsteps of their insect prey!

Although dogs can hear higher auditory frequencies than humans, they are most notable for their sense of smell. *U.S. News & World Report* recently had a story of a dog that apparently can detect the onset of its owner's epileptic seizures—45 minutes before they occur! The pet's early warnings allow the victim to prepare for the impending attack, for instance, by avoiding hazardous activities like driving. Family members say the accuracy rate is as high as 97 percent. Clairvoyance? A case of canine precognition? Probably not. It is much more likely that the dog can detect certain chemicals that may be associated with the onset of an epileptic seizure.

Descriptive accounts of these remarkable feats of sensory perception may be entertaining, but they are just the beginning of the story we wish to tell. Our ultimate goal is and ought to be an understanding of the mechanisms that underlie these abilities. What anatomical and physiological principles permit the astonishing levels of sensitivity displayed by these sensory receptors? How do these systems avoid the many sources of noise that would otherwise degrade perceptual performance? How do the animals' brains process the receptor responses, and how are these exotic modalities integrated with inputs from more conventional sensory systems? What conditions lead to the evolution of such systems, and what advantages are gained by having them? These are the questions we hope to address in the following pages.

In every case, the initial evidence for a new sensory modality came from behavioral experiments: from observations of what the animals actually *do*. The animal's behavior suggested they must possess some way of sensing environmental events that is different from our own. As early as the 18th century, the bat's ability to avoid obstacles in complete darkness was a subject of scientific investigation, although the fact that they do this by using echoes of calls they produce was not understood until the 1940s. Soon thereafter, a similar system was discovered in dolphins. The auditory modality of these animals thus has two operating modes: a passive mode by which they detect externally produced sounds, and the active, biosonar mode that relies on reflections of self-produced sonar signals.

The sensory modality of electroreception also operates in an active mode and a passive mode: some fish passively sense electrical fields produced by potential prey, while others detect prey by analyzing the disturbances in an "electric halo" that they themselves produce. We hope to do more than produce a compendium of interesting facts that modern science has discovered about a variety of odd and curious creatures, however. These exotic senses illustrate alternative ways of experiencing our planet—ways that, without science and technology, would have forever remained invisible to us. We'll get a brief glimpse of the incredible creativity and ingenuity of life on this planet. We'll celebrate that creativity, and we'll salute the ingenuity of the many people who helped unravel these marvels of the evolutionary process.

The book is a story about *what* has been discovered, and also of *how* it was discovered. And it is indeed a story, a story that has many of the elements of a good mystery. We will encounter vague and sometimes misleading clues, and find out how the true meaning of those clues was eventually deciphered by clever detectives. Often those clues were misinterpreted for hundreds of years, and those misinterpretations led to wildly unrealistic claims about what animals could and could not do. Today those old ideas seem quaint, provincial, unsophisticated, and maybe even silly. We shouldn't be too harsh or glib, however. There are no road maps to guide our search for understanding. Sometimes it is our own parochialism, our own narrow frame of reference, that provides the biggest obstacle.

How else can we explain that as late as 1912, bats were thought to navigate in complete darkness by using an elaborate sense of *touch?* How could so many learned people view the incredible, elaborate ears possessed by bats, and yet discount any role for hearing? The answer is simple. People found it difficult to believe that hearing was involved because *they* couldn't hear the bats making any sounds! And there were no *instruments* that could detect the acoustic emissions the bats were in fact producing. The critical demonstrations invariably depended upon the availability of an appropriate technology.

So we are about to begin several stories of discovery. The main characters are bats, dolphins, birds, bees, fish, and other assorted creatures. Humans play a fairly peripheral role in these tales—supporting members of the cast at best. Parts of the stories are very old. Indeed, from the perspective of the characters themselves, the plots go back millions of years. And like any good story, our characters have a motive—the most fundamental motive of all life: survival. The survival of their species, pure and simple. In their quest for survival, animals have evolved some truly astonishing abilities that we shall explore and attempt to understand.

We didn't write these tales, we are simply trying to understand the plot. In many instances, our attempts go back to the beginnings of civilization. Consider, for a moment, the case of electric fish. They represent some of the oldest living vertebrates. Our first encounter was no doubt by accident. Can you imagine the reaction, the complete and utter astonishment, that gripped the first person who touched the animal we now call a torpedo ray? It can deliver an electric shock of 350 volts! Zap! Now there's a sensory experience you'll not likely forget, but we might not characterize it as exotic. Stunning, perhaps, but not exotic.

The explanations given for the remarkable powers of the torpedo ray—now *they* were exotic. That shouldn't surprise us. What could some poor ancient mariner make of such a thing? How would it fit his philosophy of life? He didn't know anything about electricity, about how it's made, or about how it affects nerves or muscles. He certainly had no appreciation whatsoever that electricity is a form of energy essential to all life.

In an interesting and entertaining account, Chau Wu (1984) tells us of the early attempts to understand the nature of the shocks produced by the torpedo ray. The first written descriptions of the shocks delivered by the Mediterranean black torpedo came from the ancient Greeks, who knew that the shocks could cause numbness and a stuporous state in anything that touched the ray. They also knew that these effects could be transmitted through seawater, iron spears, and reeds. So they were doing primitive experiments on these creatures. They were trying to explain a curious natural phenomenon.

The generally offered explanation was that the ray emitted some microscopic venom, or a substance referred to as "effluviums of their nature," which was transmitted like darts or arrows to the victim. The soporific effects were of great interest. Galen, the greatest physician of ancient times (Greece, A.D. 130–200), theorized these effects were due to a "frigidity" of the nerves, and likened the transmissions from the ray to extreme cold. This idea that the shocks were a form of cold transmitted through water and other media held sway for over a thousand years, for it was not until the late 1700s that the electric nature of the shocks was firmly established.

The rays were considered to have significant medicinal value. The official cure for headache in first-century Rome was to place a torpedo ray on the head until the pain vanished! For gout, the prescription was to stand on a ray until the numbness reached the knees (Wu, 1984). What we now call strongly electric fish may have provided the earliest treatments for the relief of pain. Interestingly, the natives of Amazonia found similar uses for electric eels. In China, electric catfish were used to treat certain nerve disorders.

Although these early uses of electrotherapy may have done more to alleviate symptoms than to cure, the scientific studies of the torpedo ray performed during the 18th century played a prominent role in our early understanding of bioelectric phenomena in general. This spawned a branch of biological science we now call electrophysiology. Of course today we know that all major body systems—the circulatory system, hormonal systems, muscle systems, and the nervous system, are powered by "electrical batteries" that are an inherent property of living cells. The ray may not be the best cure for gout, but it has contributed mightily to medical science. Indeed, the similarities between muscle cells and the ray's electric organ, in conjunction with its unique anatomy, have made electric organs a favored model system in studies of the chemical interactions between nerves and muscles—studies that it is hoped will some day provide cures for a variety of neuromuscular diseases, such as myasthenia gravis.

Understanding some other exotic sensory systems has led to practical applications that range from the control of insect pests through the development of more sophisticated sonar and radar systems to new types of effective shark repellents. We are all familiar with new devices that have become an essential part of our everyday lives—devices that would not have been invented without essential discoveries in basic science research of all types, in all disciplines. Who knows at what stage of development the computer industry might find itself were it not for the American space program of the 1960s and 1970s? And basic science will no doubt continue to make important practical contributions. You can be assured, for example, that the U.S. Navy will continue its longtime interest in animal sonar systems. The reason is simple. The sonar system of a bat rivals that of the most sophisticated man-made devices available. And the bat does it all using a brain that is the size of your thumbnail.

Surely we still have much to learn in terms of processing efficiency and elegance from these mammals that “fly with their hands”!

The search for insect sex and alarm pheromones—chemicals that have incredible powers of attraction or dispersal—will continue as we attempt to add new, natural, nontoxic weapons to our arsenal against insect pests. Despite these potential benefits, the ultimate justification for scientific research cannot be measured in terms of national defense, the gross national product, or the Index of Economic Indicators. It comes from a deeper source. It comes from the inquiring nature of the human spirit.

Science is a process. It continues today and will continue into the uncharted future. This book tells a story that is not yet complete. While we may look at early attempts to understand nature with amusement, who knows how the views we hold today will appear in the light of another 100 years of scientific progress? Who can foretell what kinds of new marvels await discovery, what important lessons we will learn, or what kind of creature will provide the instruction?

I

Biosonar: Echoes in the Night

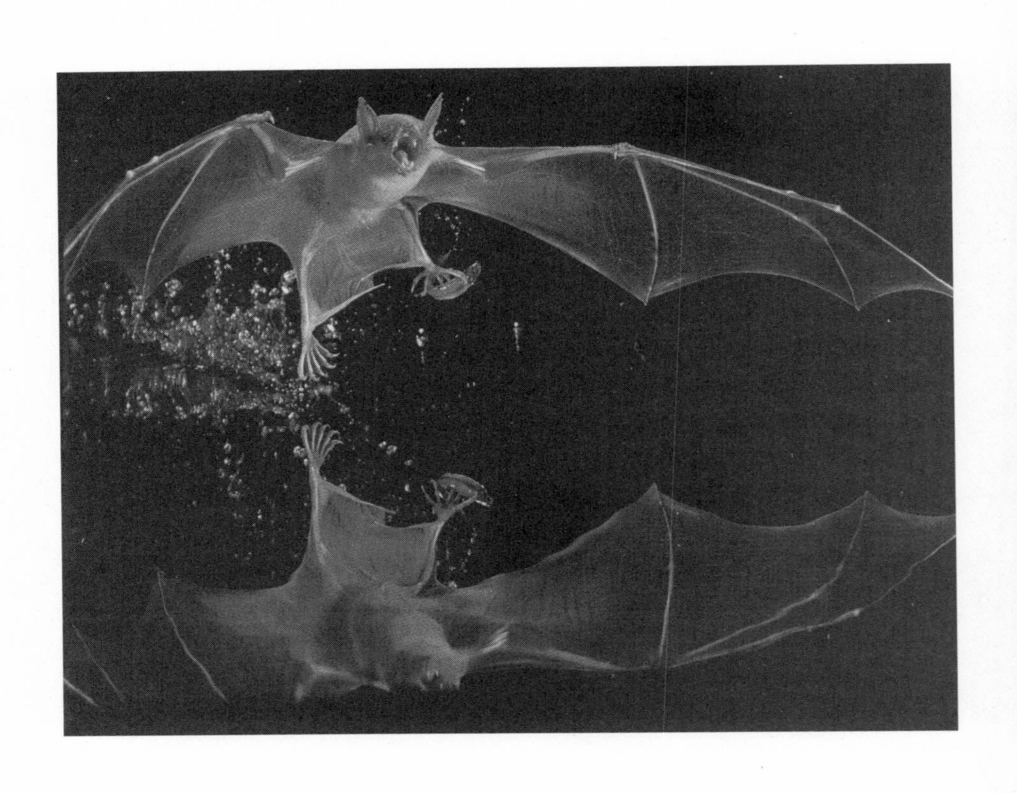

2

The Discovery

Each day, just around dusk, a truly astonishing event takes place under the rolling hills near San Antonio, Texas. At a distance, you might think you saw an enormous black cloud billowing from the depths of the earth. However, it's not a cloud of smoke that darkens the early evening sky, but the mass exodus of 20 million Mexican free-tailed bats from the depths of Bracken Cave.

Nearly every aspect of this daily event is remarkable. The bats are all female (the males live in a separate cave nearby). From first bat to last, it takes five hours to empty the cave. This may seem a long time, but imagine trying to evacuate the entire population of the greater New York metropolitan area through the Holland Tunnel in five hours! Once outside, the bats—soaring on winds of up to 60 miles per hour—attain altitudes as high as 10,000 feet. They have actually been observed by flight controllers on airport radar.

All of this is activity has one purpose. The bats are in search of a good meal. They eat insects. A lot of them—up to 1 million pounds each night. The fossil record suggests that bats have been hunting like this since at least 45 million years before the earliest hominids took their first tentative upright steps. Mothers must leave their young in the cave, and when their feeding excursion is over, each mother successfully finds her own pup among the swarming millions that were left behind.

Remarkable, don't you think? But most remarkable of all is the fact that bats can perform each of these tasks—navigation, prey detection, capture, even finding their own young—in complete darkness.

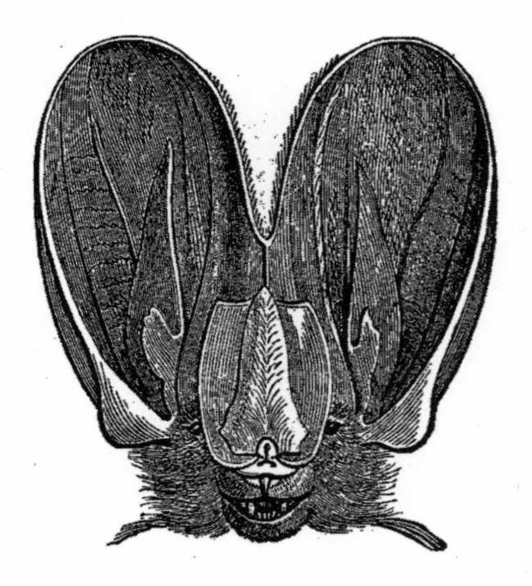

Figure 2.1
In the late eighteenth century, Professor Lazzaro Spallanzani performed his pioneering studies of biosonar on creatures such as this, the African magaderm.

The bat's ability to navigate in total darkness was known to the scientific community by 1794. The initial discovery was made by an Italian zoologist, Lazzaro Spallanzani, who was a professor at the University of Pavia. Spallanzani had been studying the ability of nocturnal owls to navigate within his laboratory under conditions of low illumination. On one occasion, the owl's wing beats extinguished the small candle that provided the only source of light. Spallanzani observed that, once shrouded in complete darkness, the owls were helpless. They were completely unable to avoid obstacles, and would collide with walls or any other object in their path.

Spallanzani then turned his attention to bats. (See figure 2.1.) He discovered that, unlike the owl, bats could successfully avoid all obstacles in complete darkness. His curiosity stirred, Spallanzani set out to determine how bats could sense their surroundings in the dark. At first, he hypothesized that perhaps the bats possessed eyes that were exceedingly sensitive, so he attempted to rule out the use of visual cues as rigorously as possible. He tried fitting the bats with small hoods that were intended to serve as a blindfold. The results were mixed, but in general the hoods did appear to interfere with their ability to navigate. Spallanzani was dissatisfied with this outcome, however. He suspected that the hoods might disrupt more than vision, and decided to test blinded bats. How did he do it? He thrust red-hot wires into their eyes. Can you imagine holding down one of these fierce little creatures while you thrust hot wires in their eyes? You have to be brave, and you have to be very quick.

In later experiments, just to be sure that the bats were truly blind, Spallanzani actually removed the eyes, using a surgical procedure that today is euphemistically called *enucleation* (Spallanzani was nothing if not thorough). To his astonishment, the flight of these blinded bats was flawless. Clearly, the navigational ability of bats did not rely on vision. But what sense did they use? It was speculated that bats had a mysterious "sixth sense."

The next year, a Swiss surgeon named Charles Jurine observed that inserting plugs into the bats' ears rendered them as helpless as owls in darkness. Jurine concluded that the hypothesized "sixth sense" was the sense of hearing. Spallanzani found this conclusion incredible, and tried to refute it with additional experiments. However, after repeated replications of Jurine's basic finding, Spallanzani was finally forced to concur. Science does work, and a good scientist will eventually abandon wrong-headed ideas when forced to do so by the weight of empirical results. Especially when they obtain the results themselves!

One reason for Spallanzani's initial skepticism that bat navigation relies on hearing was the absence of any audible sounds during their flight. Beyond that, how could sounds provide an image of the surrounding environment? The idea seemed preposterous. In fact, virtually no one (besides Jurine and Spallanzani) believed it.

One prominent scientist of that day, Baron Georges Cuvier, derided Spallanzani's work as cruel to the bats (as though that justified a summary dismissal of the results). Cuvier had a different theory. He proposed that bat navigation was somehow based on an elaborate sense of touch! In an interesting account of the history of these ideas written in 1942, the famed auditory physiologist Robert Galambos notes the sarcastic tone of Cuvier's 1795 article that criticized Spallanzani's work:

> Do not let us be surprised if a blind bat perceives the irregularities in a tunnel, the crevices and protuberances in the walls, the sticks, branches, etc.; in a word, let us not be surprised if a bat makes all the motions which have surprised the

Figure 2.2
Illustration depicting the theory that biosonar was mediated by an elaborated sense of touch. The white, "sensitive spots" on the wing are more likely blood vessels than "nerve centers."

professor at Pavia. It is simply a sense of touch developed in certain aspects to a very high degree which enables them to act as they do; it is not necessary to suppose that they possess a sixth sense.

"Conjectures sur le sixième sens qu'on a cru remarquer dons les chauves-souris, lues á la Société d'Histoire Naturelle, le 17 ventose, par G. Cuvier," *Magasin encyclopédique* 6(1795): 297–301.

It now seems incredible that Cuvier's theory, which was presented without a shred of supporting evidence, was almost universally regarded as more plausible than the Spallanzani–Jurine hypothesis, which was based on sound scientific observation. But that is what happened. For more than 100 years, Cuvier's view that the bat's sixth sense was based on a sense of touch was accepted as scientific fact. (See figure 2.2.)

Cuvier's theory was a good one in one important sense, however. The touch theory was actually testable. Theories that are not testable have no real value in science. It may seem counterintuitive, but theories that are experimentally disconfirmable are more valuable than theories that cannot be disconfirmed because they are untestable. (Sigmund Freud's theory of the human psyche is a classic example of a theory not susceptible to disconfirmation.) Not so Cuvier's theory, however. And there were experimental tests of the touch theory. In 1908, Walter Hahn reported experiments in which the wings of bats were covered with varnish or, in other experiments, Vaseline (which, by the way, had been developed in Titusville, Pennsylvania, about 20 years earlier by a New York chemist named Robert Chesebrough; he later sold the manufacturing rights to Ponds, and the Chesebrough–Ponds company to this day remains a leader in the cosmetics industry). Hahn's goal was the same as that of Spallanzani and Jurine: to try to identify the critical anatomical structure by testing the flight performance of bats deprived of that structure. Hahn reported that covering the wings in this manner had no effect.

So by 1908 it was known that (1) neither vision nor touch was required for nocturnal navigation in bats and (2) plugging the ears rendered the animals helpless. Seemingly an ironclad case. But the Cuvier theory continued to prevail, and was the dominant theory of bat navigation well into the 20th century.

In 1912, almost 120 years after Spallanzani and Jurine's work, *Scientific American* published an article entitled "The Sixth Sense of the Bat." It uncritically accepts Cuvier's explanation of the sixth sense of the bat:

> . . . the Baron Cuvier (1769–1832), was the first really to appreciate the results of the then known experiments and he arrived at the conclusion, now generally accepted, that the wonderful power possessed by the bats of directing their flight in places so dark as to render the sharpest eyes useless, was due to an exceptional development of the sense of touch, residing especially in the great delicate membranous expanse of the wings. (p. 148)

The article describes the views of Sir Hiram Maxim, a prominent engineer and prolific inventor of the late 19th and early 20th centuries. (See figure 2.3.)

Born near Sangerville, Maine, in 1840, Maxim became the chief engineer of the United States Electric Lighting Company in 1878. Three years later he immigrated to Great Britain and became a naturalized citizen of the United Kingdom. His most successful inventions came in the area of armaments. In 1884 he invented the machine gun, which used the recoil of the barrel to load and fire the next round from a canvas belt. The weapon, known as the Maxim gun, was adopted by the British army in

Head of Brainville's Bat, Which Shows the Highest Development of the Organ of the Sixth Sense to be Found Anywhere in Animated Nature. The Whole Face, Including the Ears, is Covered With This Organ; the Nose, Ears and Chin, Are All Occupied and Covered with Sensitive Hairs. The Eyes Are Small and of Very Little Use.

Figure 2.3
A drawing and its caption from Sir Hiram Maxim's article, "The Sixth Sense of the Bat," which was published in 1912. The artist may have exaggerated the ferocity of these little beasts, but not by much.

1889, but was soon eclipsed by the American-made Browning machine gun. Despite this, the Maxim gun made him a rich man, and he was knighted by Queen Victoria in 1901.

Sir Hiram apparently did not suffer from any false sense of modesty or a lack of self-confidence. He also appears to have been a man of boundless energy. Fresh from his successes with the Maxim gun, he turned his attention to the development of the airplane. He spent what at that time was a small fortune ($100,000) developing his biplane test rig. Sir Hiram was a big thinker. A really big thinker. His airplane was powered by two large steam engines, each generating 180 horsepower.

These gargantuan engines were matched by two enormous propellers whose blades were almost 18 feet long. The biplane was certainly an ambitious project, especially considering the fact that, if successful, it would have been the first machine in the world to achieve powered flight. Meanwhile, the Wright brothers were taking a different approach: their prototype was much smaller, lighter, less powerful, and, as it turned out, infinitely more successful.

A long track was constructed to guide the huge plane during testing. The track was a double-tiered arrangement in which a lower railroad track was used to guide the plane down the runway, and an upper wooden track was intended to keep the plane from achieving too great an altitude. On July 31, 1894, nine years before the Wright brothers' first successful flight, the biplane test rig lumbered down the long track. Picking up speed, it briefly became airborne, broke through the upper rail system, and crashed.

Much of Sir Hiram's interest in powered flight revolved around his conviction that it would revolutionize warfare. He was much influenced by the prophetic science fiction writings of Jules Verne, and correctly foresaw (along with Verne) the military importance of airplanes. It is likely that Maxim's fascination with flying was at least one source of his interest in bats, which are, after all, the only mammals capable of powered flight. Certainly his views on the bat's sixth sense shared at least one important similarity with his flying machine: both were bad implementations of a good idea.

Sir Hiram's mistake was to accept Baron Cuvier's notion that the sixth sense was based on touch receptors. Along with Cuvier, he disregarded the findings of Spallanzani and Jurine. Rather, he suggested that the wing beats during flight produced low-frequency sound vibrations, which Sir Hiram estimated at around 10 cycles per second (10 hertz, or Hz). Sir Hiram noted that these low-frequency vibrations " do not appeal to our ears," but could travel "after the manner of a sound" and strike all surrounding objects. The waves were reflected by these objects, and in the process were "modified by [the objects'] character and size." In accord with Cuvier's erroneous view, Maxim's contention was that these reflected echoes were detected not by the ears, but by sensitive organs of touch located in the wings and face!

Hiram Maxim's *Scientific American* article may be more fascinating today than it was in 1912. First, it provides quite an insight into the sociology of science. Being a human enterprise, science certainly has a sociology. In the history of scientific thinking, it is not uncommon to find that, either because of shared misconceptions, dogma, or appeals to authority, scientists can be misled and at least temporarily misguided. They can even maintain views that are incompatible with established facts.

In retrospect, it seems astonishing that the scientific community could so completely miss the mark on bat navigation. How could Maxim ignore Hahn's Vaseline experiments? Perhaps proponents of the touch theory would counter that the "sensitive regions" were distributed over the face as well as the wings. Did Hahn varnish the faces? Did he smear Vaseline over the elaborate nose leaves? Apparently not. Interestingly, that very manipulation may easily have impaired the bats' ability to generate the sonar calls they depend upon. In that case, it is likely that Hahn could only have concluded that his experiments confirmed the touch theory! Who can say how many misguided experiments would then have been performed?

Of course, the beauty of science is that it is self-correcting. Eventually ideas that are incompatible with observation must be abandoned in favor of those that are consistent with known facts. Science is the only enterprise that has this self-correcting property. Other systems of human beliefs or ideas, like philosophy, religion, or other moral systems, are not subject to the same criteria as scientific ideas and opinions. So it was inevitable that Cuvier's theory would crumble, because it was wrong. But bad theories can persist for some time, even when the evidence is against them.

Consider, for example, that the physical principles of diffraction (bending) and reflection of waves were understood in some detail 100 years before Sir Hiram's misguided argument, due in large part to the work of the French physicist Augustin-Jean Fresnel (1788–1827). On the basis of the work of Fresnel, the famed British physicist Lord Rayleigh (1842–1919), and others, it could have been anticipated that only the largest of objects would produce any appreciable reflection of very low-frequency sound waves. For example, the 10 Hz vibrations suggested by Sir Hiram have wavelengths of approximately 34 meters. Smaller objects would not

Figure 2.4
The "little leaves" in front of this bat's enormous ears are known as the *tragus*. They correspond to the little bump in front of the canal in our own ears. The tragus is not found in megachiropterans, and its size varies widely in the echolocating microchiropterans. It is thought to help localize the direction of returning echoes, but its function is almost surely not to "vibrate in unison with very low notes," as suggested by Maxim, 1912.

produce much of a reflection, so such a system would be of only very limited acuity. In retrospect, it seems obvious that the work of Spallanzani and Jurine, however, implied the sixth sense could easily resolve objects much smaller than 34 meters.

Beyond that, consider the presumed mechanism for detecting the reflected vibrations. Maxim's article includes wonderful drawings of a variety of bat faces (see figure 2.4). In many cases, the ears of these animals can only be described as enormous (figure 2.5). One might think it a reasonable deduction that animals equipped with such colossal ears might rely heavily on hearing, but the point of these illustrations was to show that the ears and much of the face and wings were the "sensitive portions" that enabled the elaborate sense of touch. Sometimes it's hardest to appreciate the importance of that which is most obvious!

All the same, we know now that many of Sir Hiram's ideas were valid. Certainly the idea that the "bat obtains its knowledge of the surrounding objects by sending out certain atmospheric vibrations and receiving back, in a fraction of a second later, the reflected and modified vibrations" is in principle correct. Ever the inventor, Sir Hiram describes a device

Figure 2.5
A spotted bat, one of several species with exceptionally long ears.

intended to apply the principles of the sixth sense to the avoidance of collisions at sea. The *Scientific American* article was published six months after the *Titanic* disaster. Sir Hiram's idea therefore correctly foresaw the development of sonar (*So*und *N*avigation *R*anging) during World War II.

Of course, the errors in "Sir Hiram Maxim's contention" concerned the mechanisms for the production of the sound waves and the sensory organs through which their echoes could be analyzed. As we've seen, the physics of the situation requires that the bat use very high sound frequencies, not very low ones. This is because the reflections of sound waves depend upon the relative size of the wavelength of the sound and the size of the object. The low-frequency vibrations assumed by Maxim would *only* detect objects as large as the *Titanic*, although they might have also detected the iceberg that led to its tragic sinking. Much higher sound frequencies are needed in order to detect small objects, and for this reason, both man-made sonar systems and the biosonar systems of bats use very high frequencies, ranging from 30,000 to 100,000 Hz. Rather than being too low, these sound frequencies are in fact too high to appeal to the human ear.

We now know the purpose of those enormous ears, and understand something about the elaborate "nose leaves" that so many bats possess.

This understanding began with the observations of Donald Griffin, an undergraduate student at Harvard University in the 1930s. Griffin was convinced that bats navigated through the dark mazes of their home caves by emitting very high-frequency sounds and listening to their echoes. Trouble was, he had no way to demonstrate it. When Griffin learned of a physics professor named George W. Pierce at Harvard who had developed an ultrasonic recording system, he tried to convince the professor to allow him to use it to record bat calls. The system was able to record very high sound frequencies, and electronically add the recorded signal to a high-frequency oscillator. The high-frequency oscillator was controlled by the experimenter. The summed signals oscillate at a frequency that is equal to the difference between the frequency of the bat call and the oscillator's signal. So the experimenter would, for instance, set the oscillator to 31,000 cycles/second. If the bat emitted a sonar call at 30,000 Hz, the difference between the two signals would be 1,000 Hz, which is easily audible. In this way, the researchers could detect when the bats emitted an ultrasonic sonar call. More modern techniques allow scientists to record the bat calls, then render them audible by playing the recordings back at a slower speed. The physics professor acquiesced, and the first recordings of bat sonar calls were made in 1938 by Pierce and Griffin. Later work was done in collaboration with a noted auditory physiologist, Robert Galambos. The recordings were obtained in a soundproof room. The bats wore little blindfolds made of a cellulose type of material, and the room was rigged with wires that the bats had to avoid. As the bats flew by in apparent silence, he switched on the recorder, and was rewarded with the first observations of ultrasonic bat calls. Griffin had discovered the true basis of the bat's sixth sense: biosonar. And in so doing, he had discovered an entirely new sensory modality.

There is an old expression, "When the only tool you have is a hammer, all problems look like a nail." This, it turns out, was not really the case for the prominent Dutch zoologist Sven Dijkgraaf, a man whose work we shall revisit when we consider electric fish. In the early 1940s, Dijkgraaf was working on the problem of bat navigation. Unfortunately, this was during the Nazi occupation of the Netherlands, so Dijkgraaf and Griffin were unaware of each other's work. Also, Dijkgraaf did not have access to the new ultrasonic recording technology. Nonetheless, he was able to make some important observations based on faint audible sounds the bats sometimes make. For instance, Dijkgraaf described increases in the emission rates of these faintly audible signals when bats were presented with particularly difficult navigational challenges and when they were actively pursuing insect prey. It was only after World War II ended that Dijkgraaf's work on bat navigation became widely known, and by that time, Griffin's ultrasonic recordings were clearly the preeminent data available.

In the 50-plus years since Griffin's landmark recordings, the pace of scientific progress in understanding biosonar has been remarkable indeed, despite the slow start and the red herrings introduced by those who found it impossible to accept the reality that that some animals could produce and hear sounds that were inaudible to humans. Once thought to provide only a crude representation of the world, we now know that the "acoustic image" produced by bats is astonishingly acute: many bats can detect objects as fine as a human hair by using reflected sound waves! Others turn off their sonar systems while hunting, preferring to listen passively for potential prey. It is thought, for instance, that certain desert bats find their favorite meal of centipedes and scorpions by actually hearing their footsteps on the desert floor (figure 2.6). Bats use their biosonar abilities the way other animals use their eyes: for both navigation and hunting. Many can actually identify different *types* of insects based on distinctive features of their echoes.

How does this remarkable sixth sense work? Do animals other than bats possess it? How did this ability evolve? The answers to these and many related questions serve as vivid examples of the exquisite nature of biological adaptation, and, as we shall see, illustrate the intimate relationship between biological structures and their functions. They rekindle our sense of wonder and beauty in the diversity and efficiency of life. And they reinforce our faith in the inexorable progress of the methods of science.

In terms of the variety of species and their absolute numbers, bats are the most successful mammals on Earth: there are over 1,000 known

Figure 2.6
A pallid bat attacks a centipede. It is thought that the bat can actually hear the insect's footsteps.

species. These diverse species fall into two distinct families: *megachiroptera* (large bats) and *microchiroptera* (small bats). The megachiroptera are primarily fruit eaters. They do not possess biosonar abilities. The megachiroptera often have well-developed eyes, and the size of the ears is proportional to the size of the animal; their faces are reminescent of a small dog's. Megachiropteran bats can get quite large: the flying foxes of Indonesia attain wingspans of up to six feet. Diversity is the key word, however, and the smallest megachiropterans are smaller than the largest microchiroptera.

The microchiropteran bats are almost exclusively nocturnal insectivores, although some catch fish and other vertebrate prey. These remarkable animals inhabit virtually all tropical and temperate regions of the world. Those that live in temperate regions often hibernate. As with the early videotape players, there are two basic formats for biosonar. However, while the beta format for videotape players became extinct, both biosonar formats continue to be well represented among the microchiropteran bats.

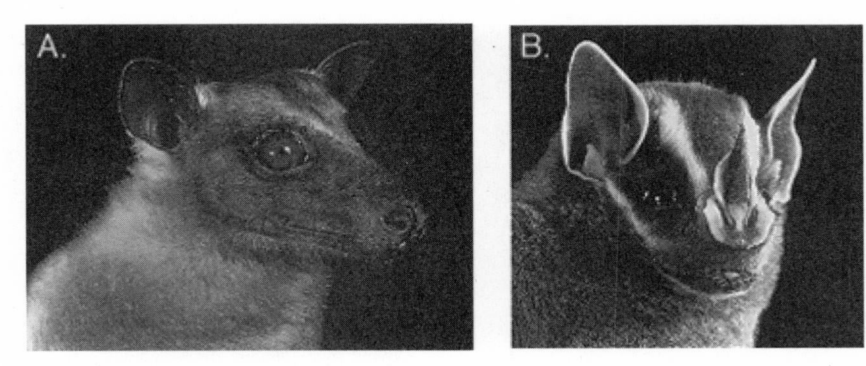

Figure 2.7
The flying fox (A), a megachiropteran species, has a doglike face. The tentmaking bat (B) is a microchiropteran species and possesses a biosonar system. Notice the tragus in front of the ear canal, and the "noseleaf," which is thought to focus the sonar beam.

One format is more prevalent in species that hunt in dense vegetation. The other is used by species that tend to hunt in open air. The two types are easily distinguished by their biosonar emissions, typically referred to as the *call*. The nature of the call in turn determines important details concerning the methods by which the echoes are analyzed. It is therefore essential that we develop some understanding of the biosonar signal.

Target
Position
Range
Approach

3

The Bat Call

From the purely physical point of view, all airborne sounds are simply variations in air pressure. These pressure variations are propagated through the atmosphere by wave motion in much the same way ripples from a stone thrown into the middle of a pond are propagated to the shore. The simplest type of sound is called a pure tone. In the case of a pure tone, air pressure changes sinusoidally with distance from the source. As illustrated in figure 3.1, this is the type of sound made by a tuning fork. Most natural sounds consist of more than one frequency, however. For example, plucking a taut string produces sound waves that consist of a combination of sine waves. The lowest-frequency sine wave is often called the fundamental. Integer multiples of the fundamental are usually present as well, and they are called harmonics of the fundamental. Increasingly more complex sounds can be produced by adding sine waves together in different combinations, as shown in figure 3.2.

Bats that hunt in heavily forested areas typically use what is called a CF-FM call. The nature of the call determines the cues available to the sonar processing system, so we must consider it in some detail. A CF-FM call consists of a sine wave (the constant frequency, or CF component), which is followed by a sweeping reduction in frequency (which is termed the "frequency-modulated," or FM component). A constant-frequency signal, followed by a downward frequency-modulated component, is shown in figure 3.3.

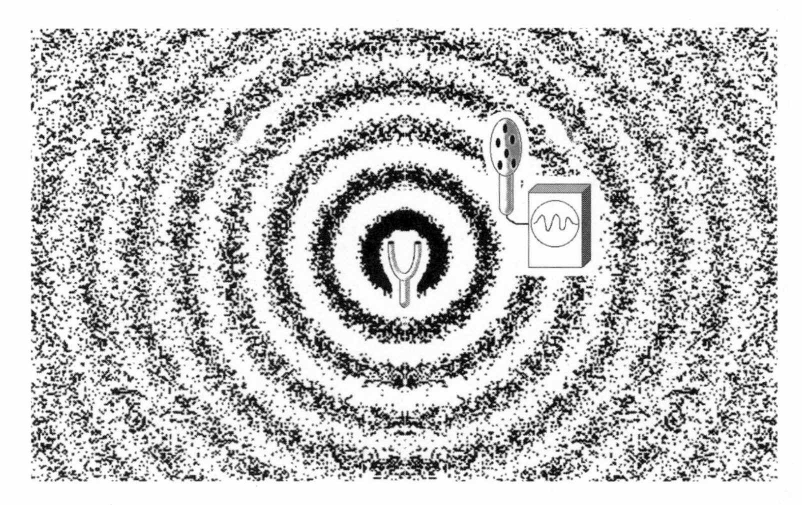

Figure 3.1
Vibrating objects produce changes in the density of air molecules. Increases in density are called condensations; decreases are called rarefactions. Here they are portrayed as variations in dot density. These variations in density produce variations in air pressure that are propagated through the air as waves. The distance covered by one cycle of the wave (one dark ring and one light ring) is the wavelength of the sound. The wavelengths decrease as the frequency of the tone increases. A microphone can sense the variations in air pressure that move past it and converts them to electrical signals that can be recorded, analyzed, and displayed. In this case, the pure tone produced by a tuning fork is displayed on an oscilloscope, which shows how the air pressure varies at the spot of the microphone over time.

Like human speech, the calls are produced in the larynx. During the call, a muscle called the epiglottis is stretched across the top of the larynx. Small amounts of air are forced past the epiglottis, causing it to vibrate at a high frequency. These vibrations produce the call. The vibrating epiglottis does not produce a pure tone. Rather, it produces sound energies at several different frequencies. As with a guitar string, there is a fundamental component and a series of harmonics. It is convenient to adopt a notation for referring to each of these components. We shall designate each component with a numerical subscript. We refer to the fundamental as the F_1 component, the second harmonic as the F_2 component, the third harmonic as the F_3 component, and so on. The fundamental frequency of most biosonar calls ranges from 20,000 to 40,000

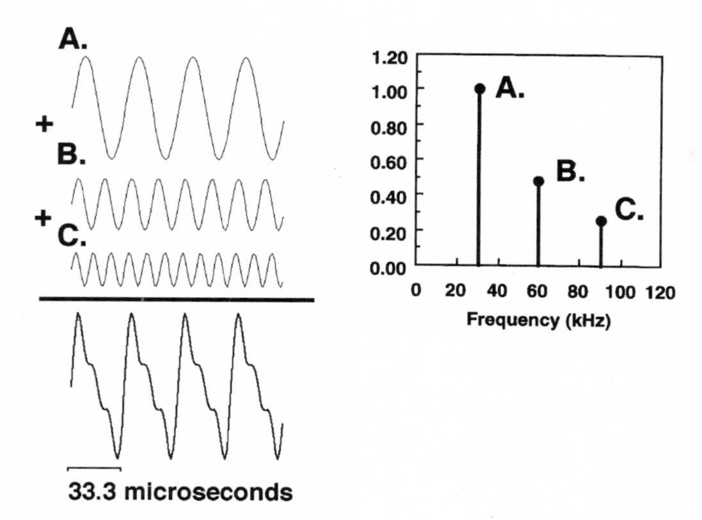

Figure 3.2
An illustration of how complex waves (including sound waves) can be contructed from sums of sine waves. In this case, A represents a sound wave of 30,000 Hz (30 kHz). B represents a sound that is twice the frequency of A (60 kHz) but has only half the amplitude. B is the second harmonic of A. C is a sound that is three times the frequency of A, so it is the third harmonic of A. When added together, they produce the sound wave illustrated at the bottom. The figure on the right is a graph of the components of the resulting complex wave. The frequencies of the different components are represented on the x axis, and their amplitudes are represented on the y axis. The three dots thus represent the components. Such a graph is called the amplitude spectrum of the wave. By convention, lines are used to connect the dots to the x axis. This makes the exact frequency components easier to see in the graph.

Hz. Since humans cannot hear sounds above 20,000 Hz, the calls are inaudible to us. The second harmonic has a frequency that is twice the fundamental frequency, the third harmonic is three times the fundamental frequency, the fourth harmonic is four times, and so on. Thus, a typical bat call might consist of an F_1 component at 30,000 Hz, an F_2 component at 60,000 Hz, an F_3 component at 90,000 Hz, and an F_4 component at 120,000 Hz. In a CF-FM bat, each of these components would be contained within the CF portion of the call, and each would be followed by a descending frequency sweep (each CF component has its own FM counterpart). In an FM bat, the call consists of a downward sweep from

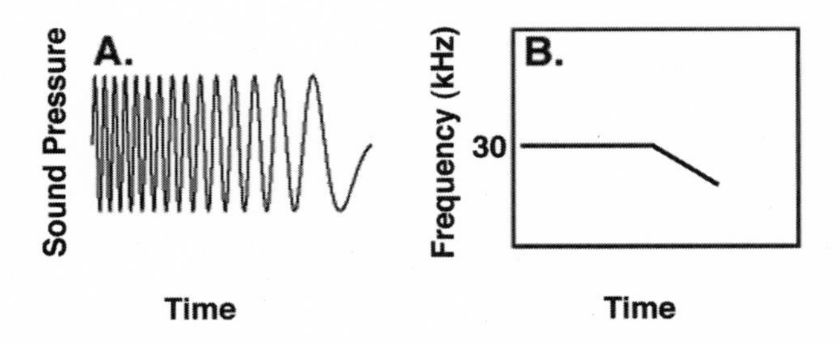

Figure 3.3
A constant-frequency signal, followed by a sweeping reduction in frequency. Since the frequency changes over time, we devise a way to illustrate three parameters of this signal: amplitude, frequency, and time (since frequency changes over time). This is accomplished using what is called a spectrogram. Here, we represent time on the x axis, and frequency on the y axis. The amplitude is usually represented by shading. Since the signal has a constant amplitude, we portray it with constant shading in this example. Spectrograms of the elements of human speech are similar to this, except that the FM component comes first, followed by the CF component.

the fundamental and the harmonics: there is no CF component. These calls are represented graphically in figure 3.4.

The graphs in figure 3.4 are known as *spectrograms*. Spectrograms are useful because they illustrate all the important features of the call. The vertical axis represents the different frequencies, and the horizontal axis represents time. The last important aspect is the amplitude of each component, which in a spectrogram is represented by the different shades of gray. Thus the horizontal segments in the CF-FM call represent the different constant-frequency components (a horizontal line means the frequency is not changing over time), and the diagonal sweeps represent the FM component (a rapid decrease in frequency over time). The CF-FM signal in figure 3.4 schematically illustrates the call of a mustache bat. There is little energy within the F_1 component of the mustache bat, so the that component is light. The F_2 component is very energetic, however, so that component is much darker.

Spectrograms are a useful way to represent any complex sound in which the frequencies vary over time. For instance, they are used to display important features of human speech sounds. Figure 3.5 illustrates a schematized spectrogram of the sound *ge* sound in the word “get.” A stylized CF–FM bat call is included for comparison. Notice the similari-

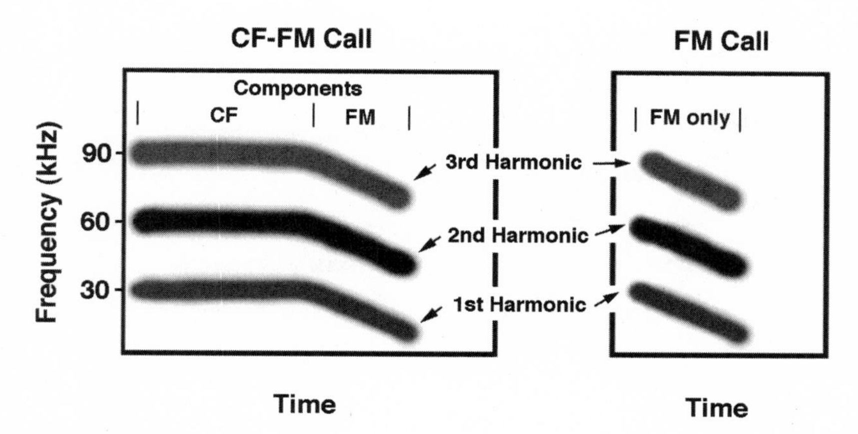

Figure 3.4
The two basic types of bat sonar calls.

ties between human speech and the sonar signal of a mustache bat. Each sound is characterized by an FM and a CF component, and each is composed of a fundamental and some harmonics. Of course, the absolute values of the frequencies are quite different, as can be readily seen by comparing the vertical axes of the two graphs. Further, the FM component in a spoken consonant usually occurs at the beginning of the sound, whereas the FM component of the bat's sonar signal occurs at the end of the sound.

Biosonar calls have some other remarkable features. First, they are incredibly loud. This may seem a strange statement, given that humans cannot actually hear the calls. However, ultrasonic recordings indicate that bat calls are about as energetic as a very loud rock band, or perhaps a large jet airplane just before landing. In fact, if we could hear these calls, they would probably hurt our ears! The high acoustic energies associated with bat calls are needed to maximize the range (or the effective distance) over which the biosonar system operates. Even at these high energies, the range of the bat's sonar system is only about 2–3 meters. If you're as agile a flier as the bat, that's apparently plenty good enough.

The high energies of these calls are all the more remarkable when you consider the size of the creature. A human couldn't possibly make vocalizations as intense as those of the diminutive bat. It requires some

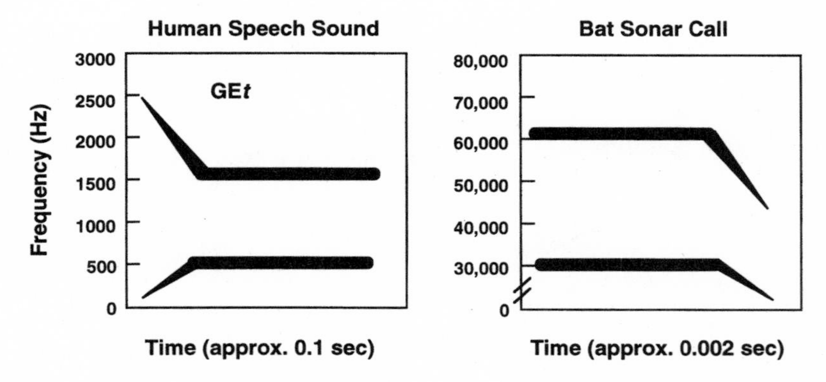

Figure 3.5
Spectrograms of a human speech sound and the sonar signal of a mustache bat.

specialized equipment. First, the bat's epiglottis becomes very tense during the call. A rigid epiglottis is needed to produce the very high frequencies the sonar system requires. A flabby epiglottis would make noise, but it would be at a much lower frequency. The sonar calls are produced by allowing air to pass this muscular membrane. However, the bat cannot let too much air pass the epiglottis because of its small lung capacity. Rapid flight using featherless membranes stretched between what are essentially the fingers of the bat's forepaws is hard work, and requires a lot of energy. The bat can't afford to devote too much of its lung capacity to making the call. So the solution to all of these engineering problems is to produce the call by vibrating a very rigid membrane. In this way, the very high, very loud frequencies that are required can be produced without using too much air.

Of course, these high-energy calls also require the generation of enormous air pressures. The pressure within the bat's larynx during a sonar call is very close to, but not greater than, the animal's blood pressure. In fact, it is crucial that the subglottal pressures not exceed the blood pressure, for if that were to happen, the call would block the circulation of blood back to the lungs. This would have a very unfortunate consequence: the bat would faint whenever it tried to produce a call.

As one might expect, given these considerations, bat calls are emitted only during the expiration phase of their respiratory cycle (try to talk while breathing in). To make their sonar calls, bats must wait to exhale.

Interestingly, the call is emitted through the nose in some bats, and through the mouth in others. The nose emission system would seem to have a certain advantage, since in this case the bat can hold prey in its mouth or eat without disabling its sonar system. Under some circumstances, bats are able to alter the characteristics of their calls in incredibly precise ways. We will learn more about that later, but for now, we simply point out that they have an ability to control the pitch of their voice in ways that would be the envy of any opera singer!

One last point about the call itself: it is obviously desirable to focus the energy of the call directly in front of the bat; dispersion of the energy tends to waste it. It's as if the bats "see" only what is illuminated with a flashlight that they must turn on and control themselves. The problem is this: the light has limited energy. If the beam is widened, more will be illuminated, but only very dimly. Better to have a focused beam that brightly illuminates what is directly in front of you. It is thought that the elaborate "nose leaves" found on so many bats serve to focus the sonar beam.

So those are the essential features of the sonar transmitter of the bat. It's a highly efficient device. One that produces astonishingly large energies at several different frequencies. A device that is equipped with specialized apparatus for focusing that energy into a narrow beam. How does the bat actually use its transmitter? Let's take a look at the bat signal in action.

Figure 3.6 illustrates the sequence of events that begins with the search for an insect, followed by detection, capture, and resumption of the search. This represents search for and capture of prey by a horseshoe bat, one of the most intensively studied species. In each "frame" we see a spectrogram of the call emitted at that point in the sequence. We show only the fundamental frequency (the F_1 component); the harmonics are omitted in the interests of clarity. We can see from the spectrograph that the horseshoe bat is a CF-FM species. The calls begin with a constant-frequency component, and typically end with a downward-sweeping FM component. Note the absence of the FM component during the search phase. Once the moth has been detected, the bat appends an FM component at the end of the call. As the bat closes in on its target, the

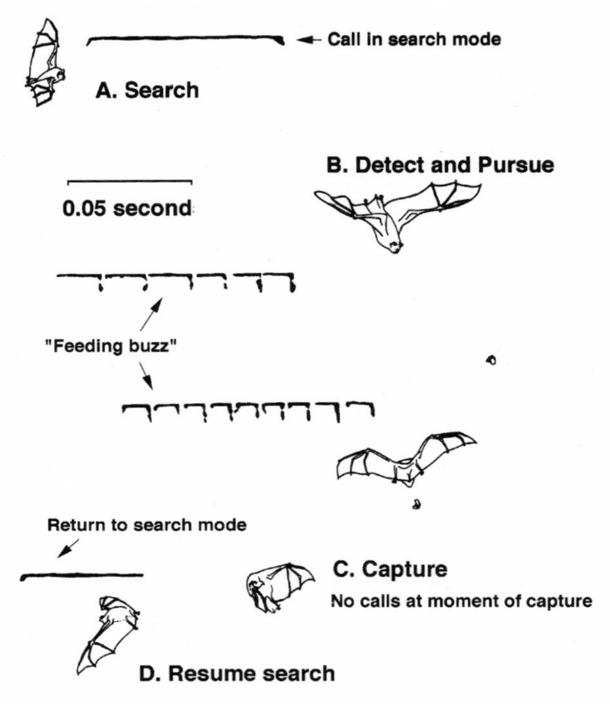

Figure 3.6
Bat sonar system in action.

duration of the CF component is reduced, and the range of the FM sweep is increased. As we've seen, the bat can emit calls only during the expiration phase. The extreme rapidity of the calls thus requires a correspondingly high respiration rate.

Why the switch to short calls separated by very brief intervals? Well, imagine that each call provides a "snapshot" image of the prey. Initially, when the moth is still at some distance, it is sufficient to take snapshots at relatively long intervals (although in point of fact, they are still only fractions of a second apart!). However, as the bat closes in, small changes in the moth's behavior become more important. The bat must be able to adjust quickly to any errors or changes that have occurred since the last snapshot, so the snapshots become more frequent; that is, the calls become shorter and the interval between calls is reduced. The very brief calls that are rapidly emitted immediately before capture are called the "feeding buzz." Ordinarily, one call is emitted during each expiration. However, in order to produce the closely spaced calls characteristic of the feeding buzz, the bats sometimes emit two calls during each expiration.

Recordings of this feeding buzz give a good indication of how frequently bats detect insects in the wild. It is estimated that bats typically detect an insect on the average of once every six seconds! Estimates as to how many of these insects are actually captured vary, since you can't tell that from the recordings of the feeding buzz alone. Most experts agree that as many as half the insects get away. This would mean that bats eat an average of 5 insects each minute, or 300 every hour. Under certain circumstances, the capture rate may go as high as 600 insects per hour!

For a brief period after capture, there are no calls. The bat is dining. Very soon the calls resume, and the animal is again in the search mode (omitting the FM component). The night life of a bat is a pretty focused affair: it's largely a matter of eating one insect after the other.

Capture techniques vary. Some bats catch the insects in their mouths. This technique obviously requires that the prey be localized with great accuracy. On other occasions, insects are captured in the membrane between the tail and hind limbs. When using this method, the bat flexes its tail and hindlegs foreward so as to make a pouch in which it catches the prey. While still in flight, it then bends its head forward and seizes the insect in its mouth. Often, if the target is out of reach of the tail, the bat catches it using a wing. It then devours its catch by bringing the wing (which is actually a modification of the forepaw) toward the mouth.

Perhaps the most spectacular method is called the somersault catch. In this maneuver, the bat flies under the prey, and after passing below it, performs a backward somersault. In the midst of this acrobatic aerial maneuver, it catches the insect in the tail pouch and eats it after completing the somersault. This entire sequence of events, from detection to capture, lasts half a second or less.

Many bats use more than one method of capture. Which one is chosen may depend to some extent on last-second movements of the insect. Clearly the wing scoop technique is less prone to error than the mouth catch. However, behavioral tests of the accuracy of the sonar system using trained bats suggest that targets can be localized with very great accuracy—perhaps within a cubic centimeter. So much of the variation may simply be a matter of preference, or perhaps reflects a kind of "hotdogging;"—showing off their remarkable athleticism! Maybe they are just having a little fun.

One is led to wonder whether the insects have any chance at all. Is this a dog fight between a Sir Hiram's biplane and an F18? Well, perhaps that is an imperfect metaphor; after all, unlike the test rig, the insects can actually fly. In any case, we've indicated that the fossil record suggests that bats have been employing sonar for at least 50 million years. The insects have certainly had the opportunity to evolve some defenses! And they have. Consider what kind of strategy the poor beleaguered insects might take.

One thing seems clear: the call itself suggests a bat is in the vicinity. After all, one can sit and listen in complete silence, but as soon as you use active sonar, you provide potential prey with an opportunity to detect *your* presence. Classic work by Kenneth Roeder demonstrated that certain moths have indeed evolved hearing mechanisms capable of detecting when a sonar beam is aimed at them. What do they do with this information? Probably just what you would do—they engage in evasive maneuvers! Certain insects fly in erratic patterns when they detect acoustic energy at about 30,000 Hz. An intriguing analysis by Altes and Anderson (1980) suggests what the optimum flight path of the moth should be. By making assumptions about (1) the velocity of the bat and the moth, (2) the turning radius of the bat and the moth, and (3) the error that the bat can tolerate and still catch the moth (their assumptions used the somersault maneuver), Altes and Anderson concluded that the moth's best strategy was to fly at a right angle to the approaching bat.

Of course, in order to do that, the insect must localize the source of the sonar beam. It does this by comparing the intensity of the beam detected by two membranes, one located on each side of its body. The bat is approaching from the side that receives the stronger of the two signals.

What if that maneuver doesn't work? What if the bug remains in the sonar beam? As the bat approaches, the beam gets increasingly more intense—something dramatic will have to be done! When the sonar beam reaches a critical intensity, the wings of the green lacewing are suddenly frozen in the upward position, and in a last ditch maneuver, it tries to drop from the sky, and to hopefully avert capture. The technique appears to work, since zoologists estimate that insects equipped with "sonar detection devices" are only half as likely to fall prey to hunting bats as those with less sophisticated equipment.

There is another defensive option. Perhaps it would be useful for the insects to generate their own high-frequency signals that might jam the sonar receivers of the bats. Insects so equipped have been discovered. They actually produce their own high-frequency vibrations. It was once assumed that this was an attempt to confuse the bat, perhaps to make it think that it was approaching a much larger object, one that might better be avoided rather than pursued. After all, the well-timed release of reflective material has for some time been a standard defensive maneuver in modern aerial and submarine warfare. The idea there is to fool the guidance system of a missile into following the material rather than pursuing your plane or submarine.

More recent thinking, however, is that, at least some of the insects that produce ultrasonic emissions are actually *advertising* their presence. Apparently, they don't taste very good, and like many poisonous or otherwise distasteful animals, they advertise their presence so as not to be mistaken for something that is more pleasing to the palate. Such animals frequently disdain camouflage. Rather, they announce their presence with bright coloration. Ultrasonic emissions of insects might be the acoustic equivalent of the poisonous frog that, by virtue of having bright red spots on an iridescent green background, ensures that all potential predators know it is better left alone.

Modern technology suggests one more strategy, but the insect world apparently hasn't hit upon it. Figure 3.7 illustrates an airplane using the so-called stealth technology. The usual shape of these aircraft is specifically designed to minimize reflections from radar devices, thus rendering the aircraft invisible to radar. The technology apparently works well for biosonar as well, since the U.S. Air Force discovered that stealth fighters left outside overnight were often littered with injured or dead bats that had apparently flown into them during the night! Naturalists have not yet discovered examples of stealth technology in the insect world, but we would expect that a "stealth moth" would be very difficult for bats to detect.

Figure 3.7
Insects might not have to worry about bats if they looked like this B2 bomber. The shape of the plane, coupled with a specialized radar-absorbing surface, produces dramatically reduced radar reflections.

So we see that the night skies are routinely punctuated with some pretty high-tech warfare between bats and the insect world. If you don't like insects, an effective (and certainly ecologically sound) method of pest control is clearly to leave the bats alone—they will reduce the number of insects in your yard. Better still, you might consider providing some bat houses in the backyard. Bats are not bad neighbors (certainly they are quiet enough), and they don't harm the environment, as many insecticides do.

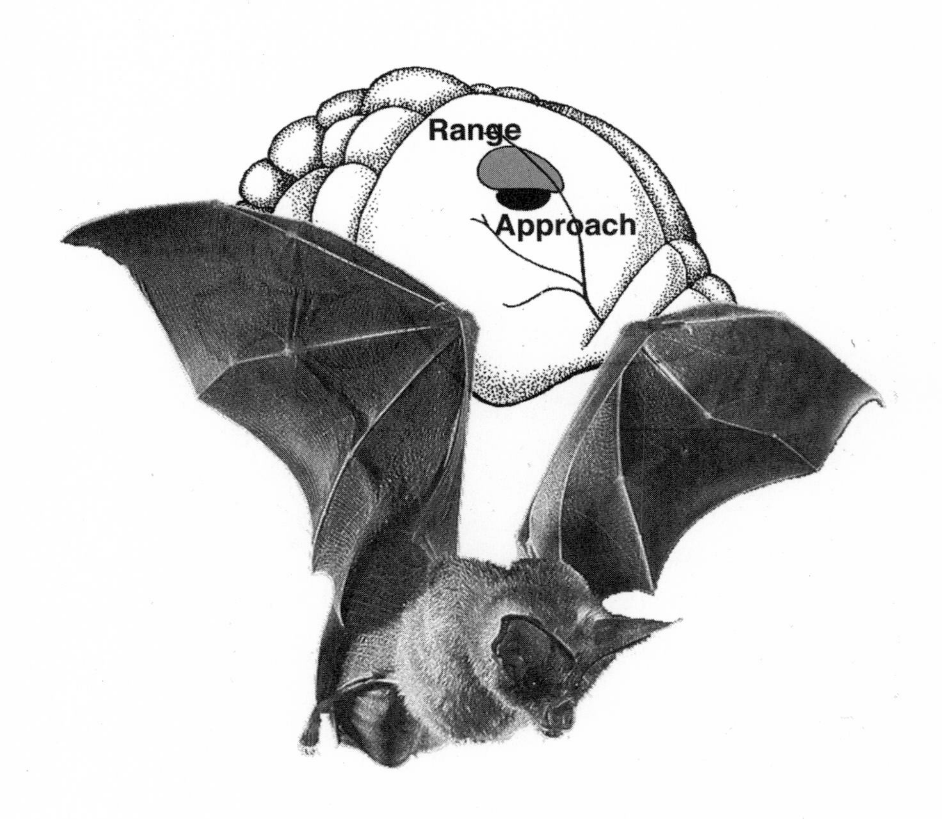
Range
Approach

4

Processing the Echo—The Sonar Receiver

To many it will appear doubtless very difficult, even on the verge of impossible, to reveal the presence of objects at sea by simply sending out atmospheric vibrations and receiving the echo of the same. One might ask, how can it be possible to judge the size, distance, and character of the object by the echo?
Sir Hiram Maxim, "The Sixth Sense of the Bat," *Scientific American*, supp. (Sept. 7, 1912): 149.

It does indeed seem difficult doesn't it? Bats *do* all these things, but they certainly don't know *how* they do it. Nor do they care. It is a uniquely human affliction to wonder how the marvels of nature actually work, and that, of course, is what we wish to learn. So we now turn our attention to the actual workings of the bat's biosonar system. We shall see how they are able to produce a remarkably detailed acoustic image of objects based entirely on their sonar capacity.

The first point we must establish is that the echo alone cannot provide unambiguous information. Creation of the acoustic image can be achieved only through a detailed comparison of the call *and* the echo. Although there are many variants, the principal issues are well illustrated by considering only one species of bat. We consider the mustache bat as a model of the biosonar system. The primary reason for choosing the mustache bat is that it is a favored species for scientific study of biosonar, especially with respect to the neural processes that support echolocation. As a result, the biosonar system of the mustache bat is particularly well understood.

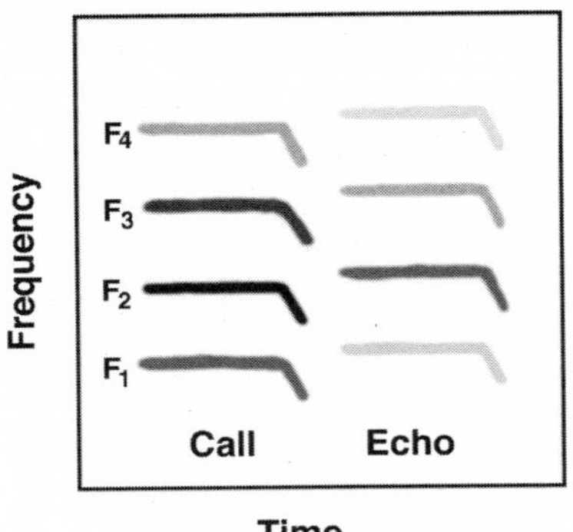

Figure 4.1
The elements of sonar processing. After a brief delay, four echoes are received, corresponding to the four harmonics of the call. Notice that the frequency of the echoes is slightly higher than the frequencies of the calls.

The mustache bat is a CF-FM bat. The fundamental frequency of its call is 30,000 Hz (or 30 kilohertz, kHz), and substantial energy is found at the second, third and fourth harmonics (F_2 = 60 kHz, F_3 = 90 kHz, and F_4 = 120 kHz). In contrast to many bats, the F1 component is not very energetic. It is the second harmonic (60 kHz) that is most energetic in the call of the mustache bat. The returning echo also has four components, corresponding to the fundamental and the three higher harmonics of the call. Figure 4.1 illustrates these essential features of the call and returning echoes. The biosonar system of the mustache bat is based on the eight elements shown: four from the call and four from the echo.

To see how different combinations of these eight essential elements can be used to produce a perceptual representation of the animal's surroundings, we first should consider what *kinds* of information the bat would find most useful. A moment's reflection suggests the following. First, the bat needs to know the distance to a given target. We refer to this distance by using the somewhat militaristic term "target range." Second, the bat needs to know how large the object is. Third, it needs to know the object's position with respect to the bat's current direction of flight. The target could be left, right, above, or below the bat's current heading. The left–right dimension is called *eccentricity,* and the up–down dimension is called *elevation.* To summarize, a useful biosonar system must provide information about target range, size, eccentricity, and elevation. It would also be quite useful to know the rate at which the bat is approaching a given target. Although it is certainly possible that the target is moving *away* from the bat, there are several reasons why it might be necessary only to carefully monitor rate of approach. First, we expect that the vast majority of the obstacles that the bat must avoid are stationary. Second, the bat flies at much higher velocities than the insects it catches. So, in general, computing the rate of approach would be very useful, but computing the rate of recession is probably less important. One final consideration. To the extent possible, we might like to glean as much information as we can concerning the shape of relatively small objects. For example, it might be relatively safe to assume that a small object in the sky is likely to be an insect in flight. However, if a bat could obtain an acoustic image of the wing motions, it would know for sure. Beyond that, different insects have different wing shapes and beat patterns, so the wing beats can provide an acoustic signature that specifies each species. It might be nice to have a system that is acute enough to use this information, for bats, like most animals, have their food preferences.

We turn now to the problem posed by Sir Hiram Maxim. How can the bat "judge the size, distance, and character of the object by the echo"? The key lies in comparisons between the specific aspects of the call and the echo. For instance, the echo obviously follows the call because the call travels through the air at a finite velocity—about 344 meters per sec. The call travels to the target at that speed, and the echo returns to the bat at that speed. Thus, the time elapsed between the call and the receipt of the echo provides the cue to target range. At a distance of 2 meters, the time interval between the beginning of the call and the returning echo will be only about 0.012 sec, or *12 milliseconds.* This is a very brief interval of time, and it is reduced in direct proportion to decreasing distance. Thus, the bat must have precise information concerning the exact timing of the call and the arrival of the echo in order to compute the call–echo delay. Soon we shall consider what kinds of neural processing are needed in order to achieve such computations. For the moment, however, we continue our description of the cues.

All other things being equal, the size of the object determines the *loudness* of the echo. I say "all other things being equal" because loudness per se is ambiguous. Any particular amplitude of echo could result from a small object that is nearby or a larger object that is farther away. However, since target range is specified by echo delay, we can disambiguate the loudness cue by taking the known target range into account. Thus, once we compute target range, we can interpret the loudness of the echo in terms of the actual size of the object.

What about the rate of approach? Can the bat get information concerning how fast it is closing in on the target? (I probably wouldn't bring it up if they couldn't.) Rate of approach is specified by what is called the Doppler shift of the constant-frequency component. What is a Doppler shift? It is an apparent change in the frequency of a tone that occurs anytime there is relative motion between a source and a detector. Of course, in our situation, both the source and the detector are the mustache bat. (See figure 4.2.)

Doppler shifts are, perhaps not surprisingly, named after Christian Johann Doppler, an Austrian physicist and mathematician who first described the effect in 1842. Doppler's principle states that the apparent frequency of sound or light increases when the source is moving toward the detector, and the apparent frequency decreases when the source is moving away from the detector. Doppler's principle explains why the apparent pitch of a train whistle increases as the train approaches the station and then decreases as the train passes. In fact, one of the earliest experimental tests of the effect was to have musicians on a flatcar play a sustained note as the train pulling the car passed a platform. Standing on the platform, a number of trained musicians with perfect pitch recorded the changes in the note as the train first approached and then receded into the distance.

Doppler shifts have played a pivotal role in modern cosmology, the scientific study of the origin of the universe. The famous American astronomer Edwin Hubble noted that light from most of the galaxies he observed was Doppler-shifted to a lower frequency (called a redshift because red light has the lowest frequency in the visible part of the electromagnetic spectrum). The observation that the light from most galaxies was shifted to a lower frequency indicated that they are moving away from our own Milky Way. The implication is that the universe is

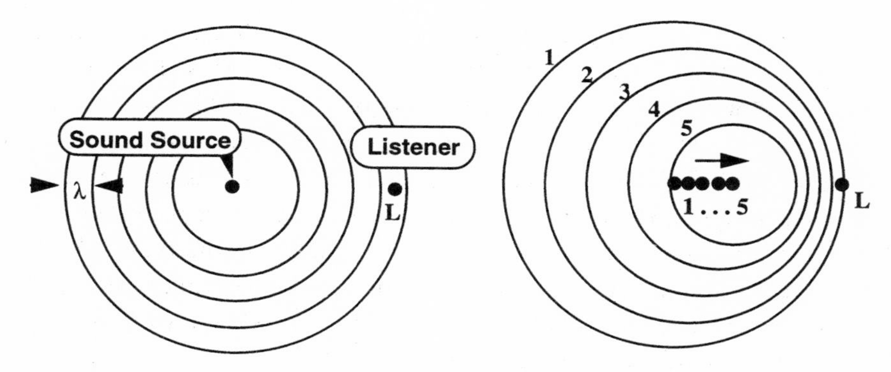

Figure 4.2
The increase in apparent frequency as a sound source moves toward a listener is called the Doppler effect. The apparent frequency will decrease as the source moves away from the listener. Each circle represents a peak in the sound pressure level. The distance between the peaks is the wavelength of the sound, and is denoted with the greek letter λ (lambda). The wavefronts "bunch up" in the direction of the movement, thus producing the apparent increase in sound frequency.

expanding, and this insight led to the development of the "big bang" theory of the origin of the universe.

How big a change in frequency is expected when a bat is closing in on a moth? For simply computing the rate of approach, we need only to apply Doppler's formula, and to do that, we need to know only the frequency of the tone and the velocity of sound in air. Given the speed at which bats can fly, we expect Doppler shifts for a 60,000 Hz sonar signal could easily be as great as 500 Hz (that is, the echo comes back at 60,500 Hz). Behavioral experiments using trained bats suggest that they can detect Doppler shifts much smaller than that. Indeed, they can detect changes in a 60 kHz signal as small as 6 Hz!

Again we note that Doppler shifts are the difference between the frequency of the energy emitted at the source and the frequency received by a detector, so their analysis requires knowledge of the exact frequency of the source and that of the echo. As we shall see, the bat's sonar receiver computes Doppler shifts.

Information about the eccentricity of the target (left–right position relative to the current heading of the bat) is obtained by comparing the intensity of the echo arriving at each ear. If the object is directly in front of the bat, the intensity of the echo arriving at each ear is the same. However, if the object is off to the left, then the left ear is closer to the source of the echo than is the right ear. The animal's head lies in the path between the echo and the right ear, so the head creates a sound shadow and the right ear falls within this shadow. For this reason, the echo arriving at the right ear is slightly softer than that arriving at the left ear. The situation is reversed for an echo originating from the right side of the animal. The magnitude of these differences in intensity increases with increasing eccentricity of the source, reaching the maximum with sources that are located directly at the animal's left or right ear.

The details of the physical basis of sound localization were described by Lord Rayleigh in a two-volume work, *The Theory of Sound*, published in 1877 and 1878. A prolific scientist, Lord Rayleigh made fundamental contributions to the physics of wave motion, optics, color, and electricity. Among his many accomplishments, Lord Rayleigh was the first to explain why the sky appears blue. Collaborating with Sir William Ramsay, he discovered the inert element argon in 1895, and for that discovery he won a Nobel Prize in 1904.

Comparing the signals arriving at each ear is universally employed within the animal kingdom to permit localization of sounds in space. Stereophonic recordings utilize the same process, in that they provide slightly different versions of a recording that are played through two speakers, one placed on each side of the listener. Binaural processing of the sounds emitted from both speakers results in the illusion that the different musical instruments are located at different positions in front of the listener; that is to say, in our terms, at different eccentricities. So there is nothing particularly unique about this aspect of the bat's sonar system. Since these localization processes require binaural comparisons, the neural systems that perform auditory localization are characterized by convergence of the inputs from each ear. Many neurons within the auditory pathways respond to sounds presented to either ear, and show specific patterns of excitation/inhibition for each ear that form the basis for subjective localization of sound sources. The bat's auditory system is actually pretty ordinary in this respect, although there are certainly some adaptations that result from the bat's need to process such high-frequency sounds.

The processing of elevation is somewhat more complicated and much less completely understood. The shape of the external portion of the ear is known to be quite important. For example, filling the little crevices of the ear results in impaired sound localization, which demonstrates that the external ear (the pinna) is more than a place to hang your jewelry! Even a casual inspection of the ears of microchiropteran bats can hardly fail to impress, and it is surely the case that the unusual ear shape is an adaptation to some requirement of the biosonar system. One suggestion is that the tragus, the appendage often located in front of the pinna, is important in producing cues useful for sensing the elevation of targets.

A Biosonar Receiver: The Auditory System of the Mustache Bat

It's said that the ears of mammals evolved from the lateral line organ of fish. That would be hard to determine for sure. But whether it's true or not, these structures have certain similarities, and we shall need to learn at least a little bit about each one.

The lateral line organ is a complicated thing. It runs down the side of the fish, just about down the middle. You know that transition that fish have between the coloration of the top and bottom, that little border? That border contains the lateral line. In some fish it spreads out pretty extensively when it gets to the face. The lateral line is part of a sensory system. Part of several different sensory systems, really. One thing it senses is disturbances in the surrounding water. Vibrations of objects in the water create variations in water pressure that change over time, and these are propagated through the water. That is essentially what a sound is, so one thing the lateral line does is sense sound waves in water. It also does some other things that are really quite remarkable, but that is best left for a later part of the book. For now, we can focus primarily on the similarities between the lateral line organ and the ears of terrestrial animals.

Anatomically, there is really only one similarity, but it's an important one. Many of the receptor cells in the lateral line organ are hair cells, and very similar receptor cells are found in your inner ear. Those hair cells in your inner hear perform the important function of converting the variations in air pressure that we call sound into bioelectric currents, which is the currency of the central nervous system. All sensory receptor cells convert some form of energy present within the environment into bioelectric currents. That's what they are built for. And a surprising number of sensory receptor cells have hairs. Besides those in the lateral line organ of fish and the cochlea of your inner ear, the receptor cells for the sense of balance, also located within the inner ear, are hair cells. In fact, visual photoreceptors of the eye are modified hair cells, too. But like the lateral line organ of the fish, we'd best leave that for a later section of the book.

In the fish, the hairs from these cells actually extend into the water. They have a little mucus cap on them, but other than that, they protrude right into the water. When the water pressure changes (from currents, sound, or the swimming motion of nearby fish), it bends the hairs, causing bioelectric currents to flow. These currents represent the nervous system's response to a mechanical disturbance in the water. That's what causes all the fish in an aquarium tank to dart away as soon as your child taps on the glass (even though the sign specifically asks them not to).

In the inner ear of terrestrial animals (including humans and bats), the situation is a little different, although the hairs still protrude into water. Of course, it's not seawater, but its ionic composition is not too dramatically different from seawater. The fluid that bathes our hair cells is called endolymph fluid, and, like seawater, it consists of salts dissolved in water. The water is filtered blood plasma, the liquid part of the blood. In any case, the hair cells, which are bathed in endolymph fluid, are located within a cavity in the bone of the skull. That cavity is called the *cochlea.* The cochleas of all terrestrial animals are very similar in shape. They all look pretty much like the one pictured in figure 4.3.

The cochlea is shaped much like a snail; it spirals upward as it turns. The number of turns varies in different creatures. Humans have about 2.5 turns. I think bats have about the same, but maybe they have more. I could check, but the number of turns doesn't seem that essential.

Here is something that is essential. Sounds travel in air, and the hairs of the auditory receptors are in water. This presents an engineering problem. The problem is this. Anytime sound vibrations attempt to move from one medium to another (for instance, from air to water), some of

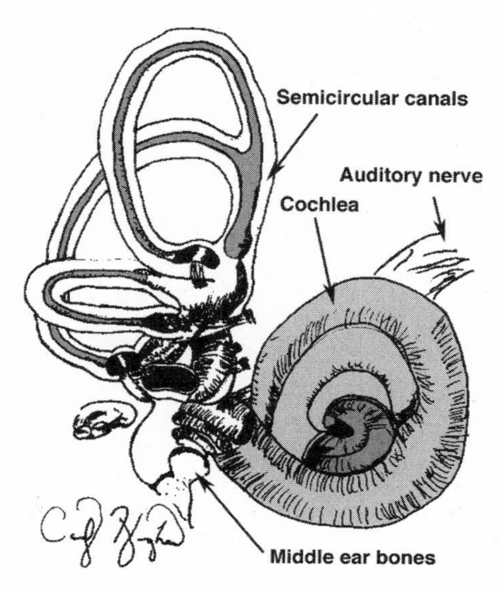

Figure 4.3
The structure of the inner ear of a typical mammal. The 3 rings at the upper left are part of the vestibular apparatus, which is involved in our sense of balance. The lower coiled structure is the cochlea. It is coiled like a snail shell, and the portion near the apex of the coil has been cut away, revealing a glimpse of the interior.

the sound is transmitted into the second medium, but a lot of it is *reflected* into the first. The question is how much energy is transmitted, and how much is reflected. Certainly we don't want too much of the energy of airborne sound waves to encounter the endolymph fluid of the cochlea, then bounce right back off. We need to transmit the energy in airborne sound into the cochlea, so that energy can propagate through the endolymph and provide the forces that bend the hairs on the receptor cells that lie within.

Without going into details, we have this problem, and it revolves around an important acoustical property of any conducting medium, the *acoustic impedance.* It turns out that the proportion of energy reflected off the interface between the air and the water is determined by the acoustical impedance of the air and water. If they are very different, then most energy is reflected. Significant amounts of energy are transmitted from one medium to another only when their acoustical impedances are

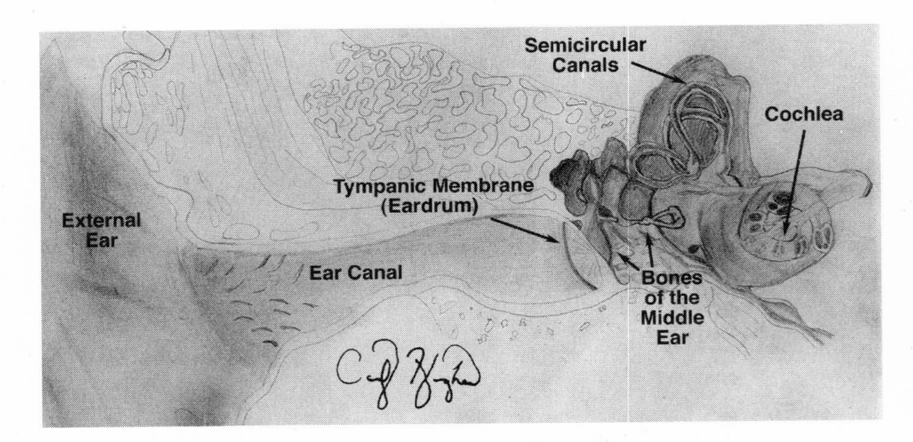

Figure 4.4
Illustration of the external, middle, and inner ear, showing the mechanical linkage of the eardrum to the cochlea via the three bones of the middle ear.

very similar. This is called impedance matching. If the values are very different, we have an impedance mismatch, and very little energy is transmitted across the interface. The impedance of air is about 40 dynes sec/cm^3, whereas that of water is 161,000 dynes sec/cm^3. The fact that these values are so different tells us that an impedance mismatch will prevent most energy from airborne sound waves from entering water. In fact, about 99.9% of the energy in an airborne sound wave is reflected off an interface between air and water. Only 0.1% is transmitted into the water. Somehow, we have to get more of that energy transmitted into the water.

Getting more energy transmitted from air into water is the job of the smallest bones in the body: the three bones of the middle ear. They're colloquially known as the hammer, anvil, and stirrup; their scientific names are the malleus, incus, and stapes. Their arrangement within the ear is shown in figure 4.4. Referring to this diagram, we see that variations in air pressure cause the tympanic membrane (eardrum) to move inward and outward. This motion causes the chain of bones in the middle ear to vibrate, and since the stapes is attached to the cochlea, vibration of the stapes is transmitted into the cochlea. This chain of three bones is very well designed. The bones of the middle ear almost perfectly compensate for the energy that would otherwise be lost because of the impedance mismatch between air and water. They do this by a combination of physical processes. First, they produce a very large mechanical advantage because they act like levers. Second, they transmit the pressure wave impinging on a large membrane (the tympanic membrane) to a much smaller membrane (the oval window of the cochlea). This produces much more force at the cochlea than would otherwise occur, and that additional force helps compensate for the impedance mismatch.

Fine. So now we have a sound wave in the fluid-filled cochlea. How is that energy going to be used to deflect hairs on sensory receptors, and how can the complex properties of the original sound wave be represented by the patterns of these deflections?

Figure 4.5 shows a cross-sectional view of the inside of the snail-shaped cochlea. It looks pretty complicated in there, and it is. But the essential point is that pressure waves entering the cochlea produce movements of the *basilar membrane* (shown in figure 4.5B). The hair cells sit on top of the basilar membrane, and their hairs are embedded in the overlying *tectorial membrane*. It is through this arrangement of the basilar membrane and the tectorial membrane that sound waves can bend the hairs of the receptor cells and begin the process we call hearing. Pretty far cry from the lateral organ of fish, isn't it?

As the stapes moves in and out, vibrations in the fluid cause the basilar membrane to move up and down. The details of this motion are complicated, but the net result is that, because they have different pivot points, the basilar membrane and the tectorial membrane slide with respect to one another, and this causes a shearing action on the hairs of the auditory receptor cells. The bioelectric currents that result from these deflections of the hairs induce impulses in the auditory nerve, which are then transmitted to the brain.

The motion of the basilar membrane is truly remarkable. Its movements change according to the frequencies of sound vibrations, so a great deal of the mechanism of hearing different frequencies relates to the way this amazing little membrane moves in response to sound waves that enter the cochlea. The details of these motions were first described by Georg von Békésy, who was awarded the Nobel Prize in medicine for his work in 1961. Békésy's experiments were ingenious and a technical tour

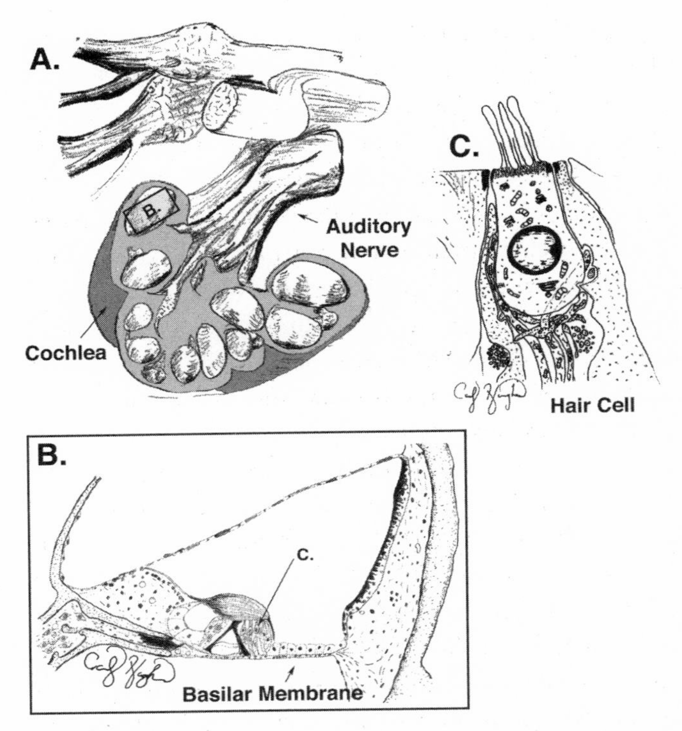

Figure 4.5
Internal anatomy of the mammalian cochlea. A. Cross section through the cochlea, showing the coiled chambers that lie within. B. Detailed drawing of the interior section indicated by the square in A., showing the basilar membrane and, lying above it, the hair cells that are the receptor cells for the sense of hearing. C. An illustration of a single hair cell. As described in the text, motion of the basilar membrane produces a shearing action on the hairs, which in turn produces ionic currents within the hair cell and, ultimately, nerve impulses in the fibers that innervate each hair cell.

de force. Using human cadavers, Békésy observed the motion of the basilar membrane using a stereo microscope, which permits visualization of the three-dimensional structure of an object.

Many formidable problems had to be solved before Békésy could actually watch the basilar membrane in action. The cochlea is embedded in the temporal bone of the skull, one of the hardest bones in the human body. So Békésy had to grind away the bone, under water, to expose the membrane. Highly reflective silver powder was sprinkled on the basilar membrane to make it easier to see. Vibrations were introduced into the cochlea by means of an electromechanical device placed where the middle ear bones would normally lie. The final problem was that the membrane moves very quickly. In fact, it moves up and down at the same frequency as the sound wave. Thus, for a sound wave of 1000 cycles/sec (Hz), the membrane moves up and down 1000 times each second. These movements are much too fast to be seen with the naked eye. Humans cannot see frequencies higher than about 40 Hz. Békésy overcame this obstacle by illuminating the basilar membrane with a strobe light. If the light flashed at the same frequency as the sound vibration, a "stop action" impression of the membrane was created. Using these ingenious methods, Békésy became the first person to observe the motions of the basilar membrane in response to sound waves of different frequencies. What did he see?

Well, as already indicated, he saw that when it moves, the basilar membrane is displaced at the same frequency as the auditory stimulus. However, each portion of the basilar membrane moves the most in response to a particular tonal frequency. The part of the membrane that moves most to low frequencies is located at the top of the cochlear spiral, the helicotrema. The part that responds best to high frequencies is located at the bottom, where the middle ear bones are attached. Thus, different frequencies cause different portions of the membrane to vibrate: the *frequency of the acoustic stimulus is converted by the mechanical properties of the membrane into a place of excitation along the length of the membrane.* This aspect of the motion of the basilar membrane corresponds to a theory proposed almost seventy years earlier by the great physicist/physiologist/psychologist Hermann von Helmhotz. In a manner not all that different from that envisioned by Helmholtz in 1870s, hearing different frequencies is a matter of determining which locations along the basilar membrane are moving. Given a little poetic license, maybe Baron Cuvier and Sir Hiram Maxim were correct: the sonar vibrations *are* sensed via an elaborate sense of touch! Of course, the receptors are located in the cochlea of the inner ear, not in the "great membranous expanse of the wings."

The mechanism for hearing in bats is not that different from that in humans. The main difference is that the bats can hear much higher frequencies, and this difference is reflected in the anatomy of the cochlea and especially the basilar membrane. If we were to uncoil the spiraled cochlea of a CF-FM bat and lay the basilar membrane flat upon a table, we would see that the width of the membrane (as well as its thickness) varies greatly as we move along its length. The part of the membrane that responds to low frequencies is wider and thinner. That which responds best to high frequencies is narrow and thicker: the high-frequency end has to be stiffer in order to respond to the very high frequencies of the harmonics of the sonar echoes.

So the different frequencies in bat sonar calls and their returning echoes are represented by the place of vibration along the length of the basilar membrane. And this pattern of vibration must be transferred into patterns of neural activity. The pattern of impulses in the auditory nerve represents a code. All the information concerning the nature of any sound must be transmitted to the brain via these fibers; any information that is not represented by this coded activity is forever lost to the listener. Therefore, sensory physiologists are in a sense cryptographers. The auditory nerve employs an elaborate code, and we wish to break that code so we can understand the neural processing that enables the bat to catch a bug, or that produces the sensation of hearing the ocean in a seashell, or that allows you to effortlessly recognize the different instruments in a Beethoven symphony. One way scientists attempt to break the code is by listening in on individual nerve cells (eavesdropping, if you will) while they are responding to carefully controlled auditory signals.

The cochlea is equipped with about 18,000 hair cells, which are aligned in rows along the length of the membrane. As a result of this arrangement, each hair cell tends to respond only to a narrow range of frequencies. Moreover, since individual nerve fibers in the auditory nerve contact only a small number of neighboring hairs cells, we might expect that the fibers themselves respond to a narrow range of frequencies. How might we demonstrate this important feature of the neural code for frequency in auditory processing?

Consider an experiment in which we are able to record the activity of a single auditory fiber. Such experiments were first performed on motor nerves by the British physiologist Lord Adrian and his collaborators in the late 1920s. Today, recordings from individual nerve cells or fibers are performed in laboratories throughout the world. Modern techniques make use of very fine wire electrodes whose tips may only be 10 millionths of a meter or less in diameter. The American physiologists George Ling and Ralph Gerard are generally credited with the first use of microelectrodes in 1949, so this is not an ancient discipline. In any case, assume that we have isolated the activity of an auditory fiber, and that we now present the animal that possesses this fiber with tones of different frequencies.

One thing we might want to know is how sensitive these fibers are to weak sound vibrations. Do they match the animal's threshold for hearing the same signals? If so, we might conclude that the animal can hear a sound whenever any one of its 30,000 auditory nerve fibers produces even a slight change in activity. (To a first approximation, this is correct!) Now, consider a case in which we find the frequency to which a given fiber is most sensitive, and then systematically change the tonal frequency, in order to determine the range of frequencies to which the fiber responds. We typically find that a given fiber will indeed respond to a tone frequency that is different from its best frequency, but to obtain the response, we must increase the intensity of the tone.

A graphical representation of the results of such an experiment is shown in figure 4.6. The tonal frequencies are represented along the horizontal axis, and the intensities of the tones are plotted on the vertical axis. The V-like shape of the graph is typical of the auditory fibers of any species. It is called a *tuning curve,* because it illustrates the range of frequencies that stimulate the fiber, the range of frequencies to which the fiber is "tuned." Notice that there is one frequency to which the fiber is most sensitive. This is called the *characteristic frequency* for that fiber. The fiber will respond to frequencies above or below the characteristic

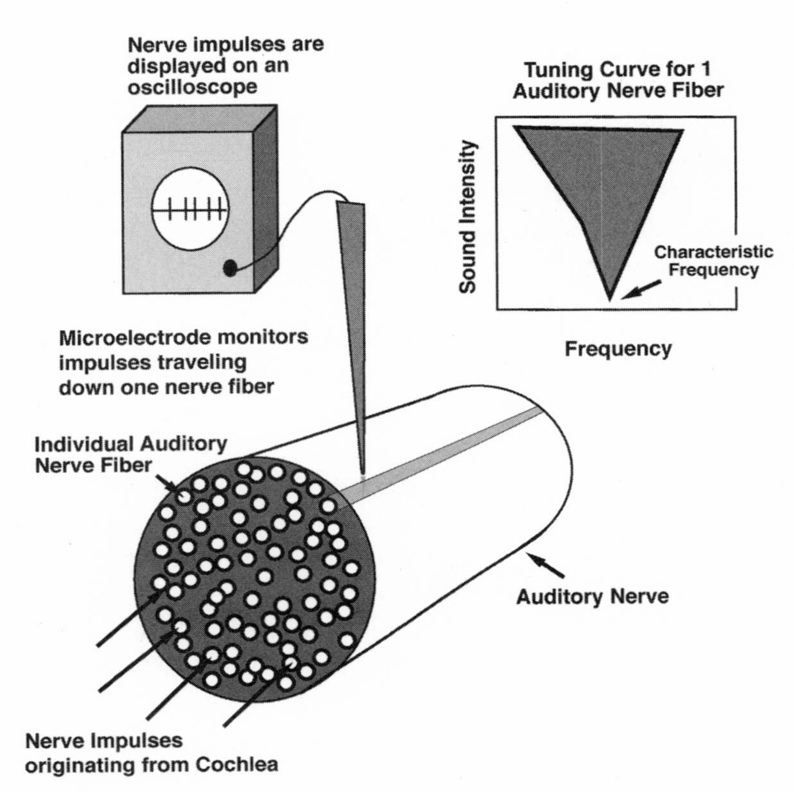

Figure 4.6
Microelectrode recording of the activity of a single fiber in the auditory nerve. One way to portray the results of such experiments is the tuning curve illustrated at the upper right. Tuning curves indicate that individual nerves respond to a relatively narrow range of sound frequencies. The set of frequency-intensity combinations that stimulate this nerve are indicated by the shaded region. The range of effective frequencies increases with increasing sound intensities.

frequency, but more intense tones are needed to produce the response. In sensory physiology, we say that the response of the fibers is frequency-specific.

Other fibers are tuned to different frequencies. As the frequency of a sound changes, there is a corresponding change in the fibers that respond to it. Thus, knowing which fibers are active is an important part of the *neural code* for the frequency of a sound. The nervous system preserves this frequency-specific pattern of activity with remarkable fidelity. For example, under ideal conditions, a bat or a human can detect a change in frequency of as little as 3 cycles/sec. So an important part of hearing relies on the frequency tuning of the neurons that process auditory information.

The degree of tuning varies from one fiber to the next, and from one species to the next. Given the requirements of a precise analysis of high-frequency echoes required by a biosonar system, it should not surprise us to find that the tuning characteristics of the fibers within the auditory nerve of echolocating bats achieve the greatest precision known in the animal kingdom. The width of the V in a tuning curve clearly reflects the degree of tuning: the narrower the V, the greater the tuning.

The tuning curves for a population of auditory nerve fibers in a mustache bat are illustrated in figure 4.7. Several things are readily apparent. First, the width of the tuning curves depends very much on the characteristic frequency of the fiber. The narrowest tuning curves are found in fibers tuned to 30, 60, and 90 kHz, which correspond to the fundamental, the second harmonic, and the third harmonic of the mustache bat's biosonar call. The auditory system of the mustache bat is clearly specialized to process echoes from the call. We're starting to detect a pattern here: the basilar membrane is particularly responsive to the frequencies in the call, and the auditory nerve contains fibers that are very narrowly tuned to those same frequencies, which are likely to be the principal components of the echo. There is only one exception: while the call does contain a fourth harmonic at 120 kHz, it does not appear to have as precise a neural representation as the second and third harmonics do. This last point is not illustrated in figure 4.7.

The auditory nerve fibers innervate neurons within the bat's brain. The nerve fibers project to the brain in an incredibly precise pattern. Auditory

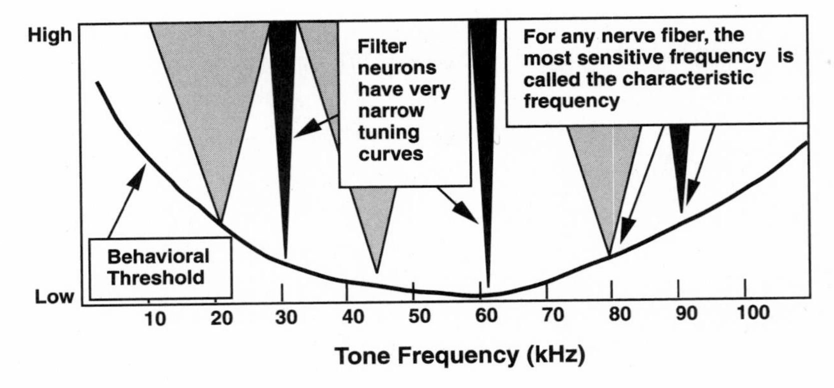

Figure 4.7
Schematic illustration of several tuning curves for different auditory nerve fibers in the mustache bat. This figure makes several important points. First, we see how different fibers respond to different frequencies of sounds. Second, the precision of the tuning depends on the characteristic frequency of the fiber. Nerve fibers that are tuned to the CF components of the sonar signal have the sharpest frequency tuning. Thus, there are populations of neurons that are precisely and specifically sensitive to the major component frequencies in the sonar signal. The curved line at the bottom of the figure represents the behavioral threshold of the bat to different sound frequencies. The bat is most sensitive to tones of 60 kHz, the second harmonic of the bat's sonar call. Also, the behavioral sensitivity of the bat is closely matched to the most sensitive auditory nerve fibers. This implies that that bat hears a tone whenever the tone is loud enough to stimulate any small subset of its 30,000 auditory nerve fibers.

nerve fibers that are tuned to similar frequencies in turn contact adjacent cells within the animal's central nervous system. Cells with similar "best frequencies" are typically found to lie in adjacent locations within the brain. In this way, each group of auditory processing neurons forms an orderly map of tone frequencies (which is largely equivalent to preserving a neural map of the basilar membrane). This is not unique to bats; it is a common pattern of organization in all biological auditory systems. The general rule is that different tonal frequencies are represented in different spatial locations within the auditory system of the brain. The thing that is unique about the neural maps of tone frequency in the mustache bat is the enormous amount of neural real estate that is devoted to processing 60 kHz. An example of one such map is shown in figure 4.8. This structure, which is called the inferior colliculus, is one of the major

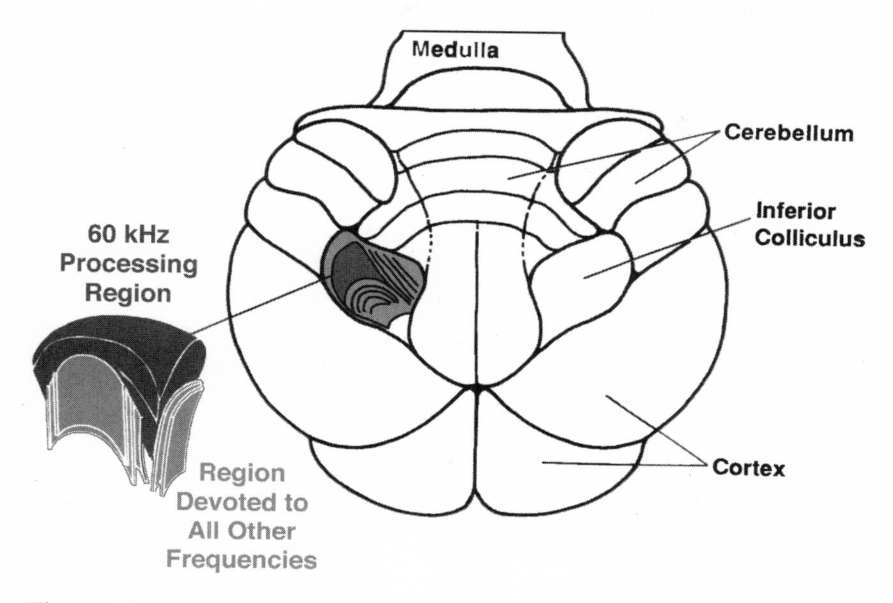

Figure 4.8
Each of the neural structures that comprise the sonar receiver of the mustache bat contains a "map" of the different frequencies the bat can hear. The map is the result of anatomical segregation of neurons according to their characteristic frequency—neurons tuned to the same frequency are clustered together within each processing center. This is a diagram of the brain of the mustache bat as viewed from above. The shaded region, the Inferior colliculus, is a major component of the sonar processing system. Within this structure, neurons that are tuned to the same frequency are arranged into "slabs" of brain tissue. The slab that is devoted to processing 60 kHz is indicated by the dark gray shading. All remaining frequencies are processed within the region indicated by the lighter shading. The 60 kHz representation has expanded dramatically relative to the slabs devoted to processing these other frequencies.

auditory processing centers in the bat's brain. At least half of the entire area is devoted to processing signals at 60 kHz. The reason for this is clear. The 60 kHz component has the greatest energy in the mustache bat's call, so the echo is loudest at 60 kHz. There is a great deal of information in this component, so the bat's nervous system naturally devotes a lot of its resources to processing signals at this frequency.

Similar principles of resource allocation occur in all sensory systems. For example, the most acute part of the human retina is the fovea. When you look directly at something, you position your eyes so that the image of the object falls directly on the fovea of each eye. It is here that we see with the greatest detail. It is here that our sense of color vision is most acute. While the fovea represents less than 1% of the total area of the retina, it takes as much as half of the resources in the visual centers of the brain. The nervous system needs to devote resources to those aspects of a sensory system that require the greatest acuity. In fact, the 60 kHz location on the basilar membrane of a mustache bat is often referred to as the acoustic fovea, a metaphor meant to highlight the fact that the bat's greatest acuity lies in the 60 kHz range. The tuning for cells most sensitive to 60 kHz is so precise that they are often referred to as filter neurons, to emphasize the fact that even a slight deviation from the best frequency renders these cells insensitive even to loud sounds.

Of course, not all of the neurons are devoted to processing the constant-frequency component of the call. Some respond to the downward sweep in frequency that follows the constant-frequency component. Those latter cells are called FM cells, whereas those which are tuned to either the fundamental or one of the harmonics of the constant-frequency component are called CF cells.

We've emphasized that extracting information about the environment requires precise comparisons between the ultrasonic calls that the bat emits and the returning echoes. Let's consider how the nervous system actually makes these comparisons. Figure 4.2 reminds us that there are four main components to the call and four corresponding components in the echo. In order to use the echo to locate objects such as insects and obstacles, the bat must somehow compute call-echo delays and Doppler shifts, and compare the characteristics of the echo as it arrives at the two ears. It must also do this very quickly. Evolution has found a number of elegant solutions to these very difficult problems of biological engineering.

Consider first the problem of computing target range by measuring the delays between the call and the returning echo. Professor Nubuo Suga of Washington University in Saint Louis has found a population of neurons within the brains of mustache bats that perform this task. These "range-finding" neurons are an example of what Dr. Suga has termed combination neurons. Combination neurons differ from neurons in the auditory nerve in that they are *specifically sensitive to particular combinations of two different frequencies*. There are many varieties of combination neurons. Some respond to two CF components, and are called CF-CF cells. Those that respond to combinations of the FM component are correspondingly called FM-FM cells. These cells are found at rather high levels in the processing hierarchy of the auditory system. It seems clear that combination cells derive their specific properties by virtue of specific patterns of convergence from lower-level auditory centers. Thus, we might surmise that a cell that selectively produces a burst of activity when presented with a tone at 30 kHz, followed by a tone at 60 kHz, receives its essential inputs from two groups of lower-level cells: one that is tuned to 30 kHz and another tuned to 60 kHz. The selectivity to the combination can be quite high. Professor Suga points out that many combination cells are several thousand times more sensitive to their preferred combination than they are to either of the components presented alone!

Where are these combination cells located? They attain their greatest elaboration in a brain structure known as the cerebral cortex. The cerebral cortex is a sheet of neurons that covers the outside part of the brain (cortex derives from the Greek word for "bark": the bark of a tree, not the bark of a dog). The cortex is generally regarded as mediating the highest levels of neural information processing. The importance of the cerebral cortex to higher mental functions is suggested by comparative studies of the organization of the nervous system. When we compare the ratio of cortical volume to total volume of the central nervous system, an orderly pattern emerges. This ratio is very small in less "intellectual" forms such as fish, amphibians, and reptiles. Mammals tend to have a well-developed cerebral cortex, and the cortex attains its greatest elaboration in the most intelligent mammals. As expected, the cerebral cortex is most developed in humans, but it is interesting to note that dolphins closely approximate the human cortical weight/total brain weight ratio. Bats lie somewhere around the mammalian average in this regard.

Each sensory modality has a representation in the cerebral cortex. Most cortical representations of the different sensory modalities have the maplike quality we've seen in other auditory centers, such as the inferior colliculus (figure 4.8). To emphasize their lower status in the processing hierarchy, these lower-order centers are called *subcortical* centers. The first-order maps in the cortex are of the sensory surface itself, so the visual cortex has a map of the retina, just as the inferior colliculus contains a map of the basilar membrane. Since the image of the world is projected onto the retina by the optics of the eyes, the visual cortex can be thought of as having a map of the visual world. Destruction of a portion of the map contained in the principal area of the visual cortex (by accident or stroke) results in a circumscribed region of blindness. The location of this blind region is entirely predictable from the location within the map that sustained the damage.

We've seen that many of the lower auditory centers have a frequency map. This map can also be viewed as being a map of the basilar membrane, since we know that different frequencies stimulate different places on the membrane.

At the lowest levels of auditory processing, simple frequency maps are the rule. Once we reach cortical levels of the auditory system, however, the neural cartography becomes somewhat more complicated. In the cortex, each sensory modality contains multiple maps. In Old World monkeys, for example, the part of the cortex devoted to vision may contain as many as 20 representations of the visual world. Multiple maps are the rule, not the exception, so we expect the cortical organization of the bat's auditory system to have many different areas. It does. It is natural to suggest that the different cortical areas perform different jobs, that each of these areas is specialized to perform specific tasks. It appears as though they do. The auditory cortex of the mustache bat provides some of the clearest indications of how the natural history and ecological requirements of a species shape the organization of its brain.

Figure 4.9 shows the location of one of the specialized sonar processing areas in the cerebral cortex of the mustache bat. Because the cells that

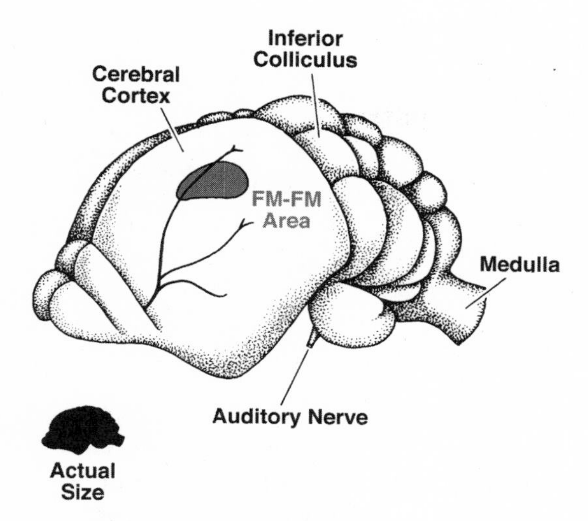

Figure 4.9
A view of the FM-FM processing area in the left hemisphere of brain of the mustache bat.

populate this area are combination cells that are tuned to FM components of the sonar signal, Suga and his collaborators called it the FM-FM area. They found three types of cells in this region of the bat's cerebral cortex. One type is responsive to a combination of the FM component of the fundamental frequency (FM_1) and the FM component of the second harmonic (FM_2). The other types respond to the fundamental combined with the third (FM_1-FM_3) or the fourth (FM_1-FM_4) harmonic. Each cell is tuned to the occurrence of its particular combination *in a specific temporal order.* Examples of this phenomenon are illustrated in figure 4.10.

We mentioned earlier that, unlike many other sounds, the sonar call of the mustache bat contains little energy at the fundamental frequency of 30 kHz. Most of the energy is in the second harmonic, at 60 kHz. Because the call has so little energy at 30 kHz, the echo at 30 kHz is very faint indeed. In fact, it is the fundamental of the call, not the echo, that stimulates the FM-FM cells. Thus, each FM-FM cell responds to the 30 kHz FM sweep of the call, followed by either the 60 or the 90 kHz sweep from the echo. In addition, each cell requires a specific temporal interval

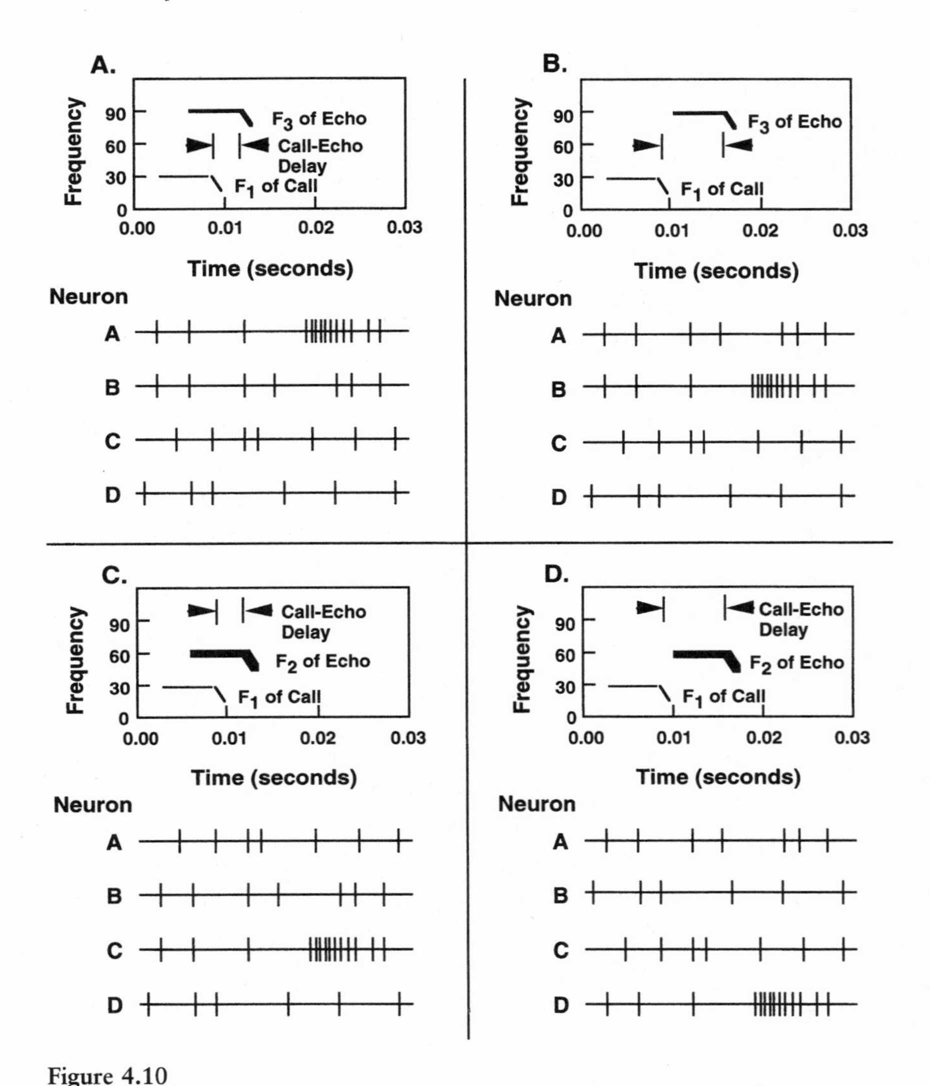

Figure 4.10
Schematic illustrations of the tuning of FM-FM neurons to call-echo delays. The responses of four different neurons to four possible combinations of call and echo are shown in panels A–D. In A, we focus on the combination of the fundamental of the call (F_1) and the F_3 component of the echo. The call echo delay is brief. Cell A is activated by this pattern, but none of the other three cells responds to this particular combination. In B, the same two components are presented, but the call-echo delay is longer, which signifies a more distant target. In this case, only neuron B responds. Neurons C and D are specific to combinations of F_1 of the call, and F_2 of the echo. Neurons C and D each respond to a different call-echo delay. Cells A and C therefore are specifically activated by a near target, whereas B and D respond selectively to a more distant target.

between these components. So we may say that the cells are tuned not only to a specific pair of frequencies but to a specific pair of frequencies that occur in a specific temporal order. This *temporal tuning* of the FM-FM cells implies that they are importantly involved in computing target range.

What type of neural circuitry can produce this selectivity in temporal delays? Well, the actual circuit has not been completely worked out. It is not possible to actually trace the exact connections between the cochlea and a specific cortical neuron. The principles of neurophysiology suggest a likely solution, however. In general, neurons respond most vigorously to two different inputs when those inputs arrive at the neuron at the same time. Since the call always precedes the echo, a useful strategy would be to arrange the circuitry such that the input associated with the call takes longer to reach the FM-FM neuron than the pathway that carries the information on the echo. In this way, the signal associated with the call gets a "head start," but the signal associated with the echo catches up because it is transmitted to the cortex faster.

There are at least two ways to create such calibrated delays in the two input pathways to the FM-FM cells (figure 4.11). One is to use a longer chain of neurons in the pathway transmitting the timing of the call. Impulses are transmitted between neurons across a small gap called the synapse. Transmission across a synapse takes some time (a minimum of about 0.5 millisecond), so the transmission time in a multisynaptic pathway is longer than in one that contains few synapses. The second method is to employ different sizes of fibers in each path. Larger fibers transmit neural impulses faster than smaller ones, so using larger fibers in the pathway that carries echo information could in principle permit the call and echo to occur at different times but arrive at the FM-FM cell simultaneously.

This strategy of creating neurons that respond to a specific combination of the call (usually the 30 kHz component) and the echo (the 60 kHz, 90 kHz, or 120 kHz component) is well suited to performing computations of target range, and can also be used to detect Doppler shifts (which represent approach velocity). It has a number of advantageous features, and is an elegant general solution to the task of comparing the echo with the call.

One might wonder why the combination cells always use the F_1 component in combination with a higher harmonic. The F_1 component is,

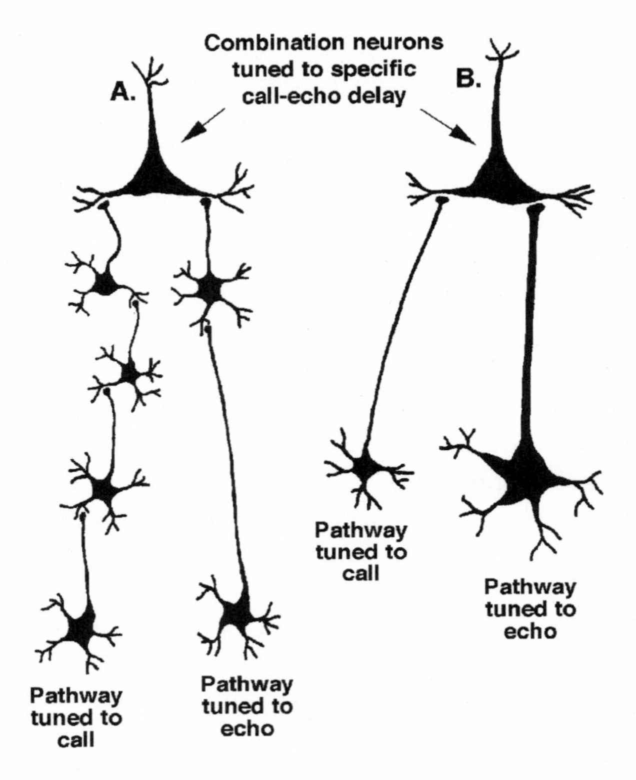

Figure 4.11
Two possible ways to produce combination neurons that are selectively responsive to a particular call-echo delay (corresponding to target range). In A, the pathway on the left conducts more slowly than the one on the right because there are more synapses along the way. In B, the left pathway is slower because small axons conduct nerve impulses more slowly than larger axons.

after all, the one with the least energy. Wouldn't it be better to use the F_2 or F_3 component? The choices made by sophisticated biological processes are usually not the result of an accident or happenstance. We should probably assume that combination cells usually use a combination of F_1 paired with some other component of the echo for a reason. What might that reason be?

We know, for example, that there are times when large numbers of bats might be flying close to each other. What would happen under such circumstances if the combination neurons relied on an alternative combination—say the FM component that follows the second harmonic of both the call and the echo (an FM_2-FM_2 cell)? Well, we might surmise that in a crowded situation, with many bats making calls with great rapidity, the FM_2 sweeps from *different bats* might arrive at temporal intervals that would stimulate FM_2-FM_2 neurons by chance alone. This, of course, would incorrectly signal targets at a certain distance. In effect, the sonar calls of other bats would provide noise that would jam or at least degrade sonar performance.

With as many as 20 million individuals living close to each other, the sonar processing system must have a way of ensuring that the combination cells are specifically responsive to the sounds that came from the same bat that made them! Use of the fundamental frequency for the first component can achieve this requirement, in part *because* of it is not that energetic. Since the fundamental is relatively weak, it is more difficult to hear the fundamental of different bats than the second harmonic. This alone increases the probability that the FM_1-FM_2 and FM_1-FM_3 cells will respond only to self-produced calls.

An interesting feature of vocal–auditory interactions also serves to ensure that the combination-sensitive neurons respond only to combinations of the bat's own call and the resulting echo. It involves the small bones of the middle ear. Two tiny muscles are attached to these bones. When they contract, they stiffen the linkages between the eardrum and the cochlea, and in so doing, decrease the sensitivity of the hair cells. These muscles contract in close association with the production of the sonar call, and thus serve to temporarily decrease the sensitivity of the auditory system in synchrony with the sonar emissions. A substantial amount of sound is transmitted from the larynx to the cochlea directly through the skull. This energy reaches the cochlea through the temporal bone, bypassing the middle ear altogether. The vibrations arriving at the cochlea through this bone conduction pathway are therefore not affected by the contractions of the muscles of the middle ear. But the calls of other bats are, so the muscle contractions help to ensure that the first component that stimulates the combination cells does indeed come from the bat's own call. Then, if a second signal arrives at the right frequency an instant later, that second sound is almost sure to be an echo of the bat's own sonar signal. An elegant mechanism, to say the least.

Incidentally, humans also have these tiny muscles that can dampen the action of the middle ear bones. When we speak, we also reflexively contract the middle ear muscles. And acoustic energy from our voices is also transmitted to the cochlea via bone conduction. That is why our voices sound so odd when we listen to a tape recording of ourselves—when we listen to the tape, the middle ear muscles are relaxed, and there is no bone conduction to the cochlea. Those differences make recordings of our voice sound strange to our ears, even though others find the recording produces an excellent likeness of our voice.

There is one last feature that helps ensure that combination cells are not "fooled," and that they respond only to the proper combination of sound vibrations. The frequency components of bats tend to differ slightly between individuals. For example, as with the human voice, the frequency of females tends to be higher than that of males. The detailed tuning of the most acute auditory neurons (the filter neurons) is well matched to the individual differences in the sonar call, providing yet another mechanism that helps the bat's auditory system tune in to its own sonar signal. That is, the filter neurons of each individual bat are precisely tuned to the exact frequency of the second harmonic of that particular individual's sonar call. The more we delve into the details of this system, the more we appreciate the elegance of its design.

Given a standard velocity for sound waves transmitted in air (it actually varies slightly, depending on air pressure and temperature), each call–echo delay is associated with a specific target range. So it is interesting to note that the distribution of neurons possessing different optimal delays (target ranges) within the FM-FM cortical area is not at all haphazard, but is integral to the map created within this tissue. Suga and his colleagues have explored this region in detail, and have noted that the cells sharing the same echo–delay preference conform to an orderly pattern within the FM-FM processing area. Figure 4.12 shows how two dimensions are represented within this tissue. One dimension is the matching or combination of the fundamental frequency of the call and the frequency of a given harmonic of the call. The other dimension is the interval between the call and the returning echo.

This map is an example of what brain scientists call computational anatomy. It suggests a how an approaching target is represented within the brain of the mustache bat. The idea works something like this. When first detected, the target is likely to be at some distance, so the call-echo

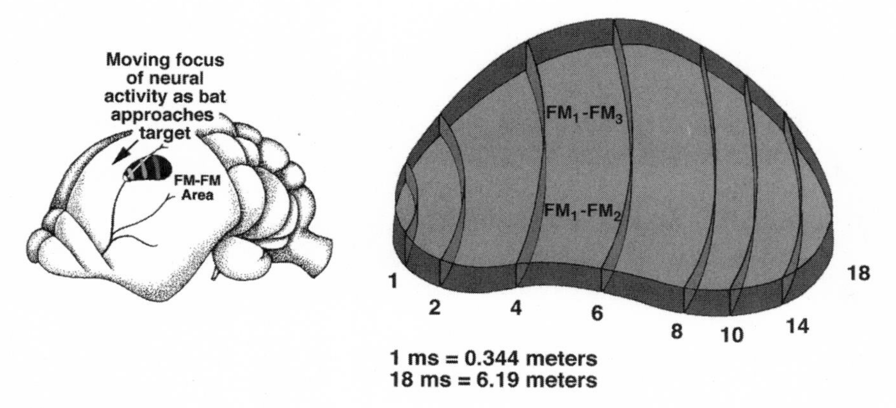

Figure 4.12
In the FM-FM area of the mustache bat's cerebral cortex, cells tuned to the same call—echo delay are arranged into parallel slabs. Longer call—echo intervals are represented toward the back of the brain, and short intervals are represented towards the front. As the bat approaches a target, we can imagine that the neural activity within the FM-FM area actually moves from back to front, as schematically illustrated in the left diagram.

delay would be relatively long. Notice how this would produce activity in combination cells at a specific location within the FM-FM processing area. For example, at a distance of 3.44 meters, the call-echo interval would be about 10 milliseconds (10/1000, or 0.01 second). The active cells would be located somewhere along the 10 millisecond echo delay line within the map. The exact location of the activity along this line is determined by the exact value of the fundamental frequency of the call. This frequency needs to be represented for reasons that have not yet been explained. For now, we simply point out that the fundamental frequency of the sonar call *changes as the bat approaches the target.* We'll soon consider why this is necessary. For now, we must simply note that if the fundamental frequency changes from 30 kHz on one call to 29.5 kHz on the next, the bat will have to possess filter neurons that respond to each frequency of the fundamental (they are so narrowly tuned to frequency, that neurons will not respond to both, so we need to represent different fundamental frequencies with different neurons). If the target is an insect, the goal is to reduce the call-echo delay to zero (capture). Thus, the bat should fly in such a way as to move the focus of activity within the

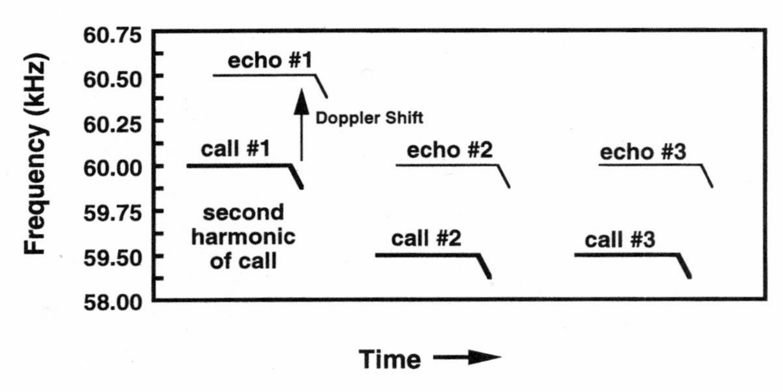

Figure 4.13
Illustration of Doppler compensation. As the bat approaches an insect, the echo from the call is Doppler-shifted to a higher frequency. The bat lowers the frequency of the next call just enough to keep the echo at 60 kHz.

FM-FM area toward the representation of 0 millisecond delays. One can imagine that if we could actually visualize it, a focus of neural activity would move forward (from right to left in the figure) as the bat approaches and ultimately captures the prey: changes in target range are represented by changes in the position of activity within the FM-FM cortical map.

An analogous arrangement produces a spatial code for approach velocity, which, as we saw earlier, is computed from Doppler shifts. However, in addition to specifying approach velocity, Doppler shifts produce certain problems for the sonar processor of the mustache bat. Let's see what the problem is, and how the bat solves it. (See figure 4.13.)

Recall that the mustache bat devotes enormous processing resources to analysis of the second harmonic of the echo, which is 60 kHz (plus or minus a few Hz for the variations that occur between individual bats). Unlike the FM-FM neurons, the detection of Doppler shifts is most efficiently performed by using the constant-frequency components of the call and the echo. The magnitude of these Doppler shifts can be quite large. For a bat approaching a target at 20 km/hour, the Doppler shift on a 60 kHz signal would be about 1000 Hz. This means that if the second harmonic of the call was exactly 60 kHz, the second harmonic of the echo would be increased to 61 kHz. And herein lies the problem.

The bat devoted all this neural machinery to processing signals at 60 kHz, and now we find the returning echo is 61 kHz! The frequency tuning of the filter neurons in the auditory nerve is so precise that they will not respond to a 61 kHz signal; Doppler shifts of this magnitude take the echo out of the range of the acoustic fovea.

Hmmm. How can the bat deal with these Doppler-shifted echoes? We could add more filter neurons, enough so that we have plenty of neurons to respond to the second harmonic of the echo, whatever its frequency. That might work, but neurons are expensive. They take up space. We'll need a bigger auditory nerve, more neurons at every level within the auditory system. We will probably need a bigger bat brain, and thus a bigger skull. The whole bat might have to be increased in size. This could be worse than we thought.

But wait! There may be a better way. What if the bat could change the frequency of the call, lower it by the amount that equals the expected Doppler shift? In this way, the second harmonic of the echo could remain right where we want it, at 60 kHz. That, it turns out, is exactly what happens. It's called Doppler compensation, and it was discovered by Dr. Hans-Ulrich Schnitzler of the University of Tübingen in Germany.

Doppler compensation can be dramatically demonstrated by harnessing a bat to a pendulum. While riding on this little bat swing, the animal is trained to make certain types of discriminations by using its sonar system. Recordings are made of the sonar calls emitted as the bat swings toward and away from a stationary target. Typical results are shown in figure 4.14. If the bat is stationary, the CF component of the second harmonic of the call remains stable at 60 kHz. However, once the bat swing is set into motion, the bat systematically reduces the frequency of the call to a degree that exactly compensates for the expected Doppler shift. This maneuver keeps the returning echoes at a constant frequency despite the fact that the Doppler shifts vary at different points along the pendulum's motion. Notice the exquisite voice control required by such accurate Doppler compensation. The bat must hear the echo, and dynamically change the frequency of its next call. This represents a level of precision in the control of the vocal apparatus that exceeds even the most demanding of operatic arias.

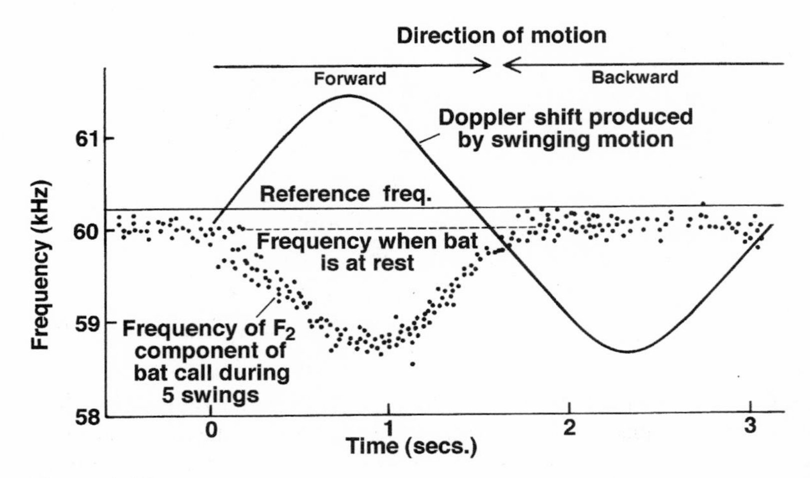

Figure 4.14
Dynamic aspects of Doppler compensation are revealed by placing an echolocating bat on a swinging pendulum.

Also note that Doppler compensation occurs only as the bat *approaches* the target. There is no compensation for the Doppler shifts that result from recession away from the target. Apparently, there is no use devoting neural resources to detailed processing of echoes reflected off objects that are receding into the distance. The bat is interested only in things that it is approaching.

Now we see why different CF-CF neurons are tuned to slightly different frequencies of the fundamental component. They have to be. Since the fundamental frequency (and, consequently, all of the harmonics) of the call changes as the rate of target approach changes, the neurons must be narrowly tuned to each possible value of the fundamental frequency of the call. However, since the Doppler compensation system works so well, the second harmonic of the echo remains relatively constant. This of course is the job of the Doppler compensation system to keep the second harmonic of the echo centered at 60 kHz, so that it can be processed by the acoustic fovea, and not by some less precise portion of the basilar membrane. Instead of producing a constant-frequency call and allowing the frequency of the echo to change with different approach velocities, the bat varies the frequency of the call in such a way as to keep the frequency of the echo constant.

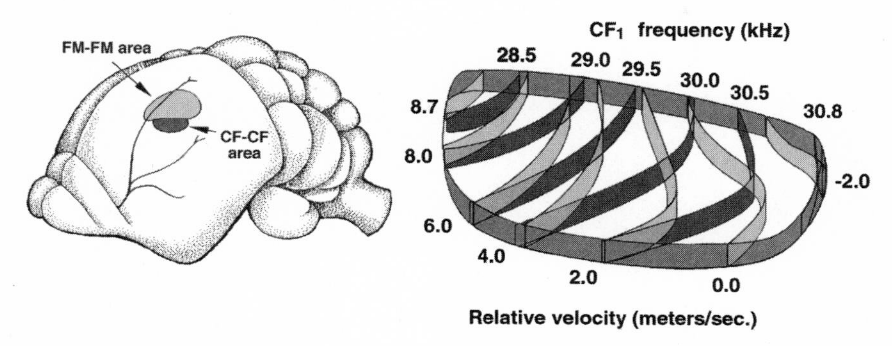

Figure 4.15
The CF-CF area is specialized to compute Doppler-shifting of the the echo relative to the sonar call. Neurons tuned to different CF_1 frequencies of the call are needed because of variations in call frequency produced by Doppler compensation. Neurons with characteristic frequencies corresponding to the same CF_1 frequency are aggregated into "ribbons" of cortical tissue (indicated by darker shading). Combination neurons tuned to the same Doppler shift are also aggregated into ribbons (lighter shading). Since Doppler shifts signify apporach velocity, the locus of activity within this area changes as the rate of approach changes.

A match must still be made between the frequency of the call and that of the echo. This information is critical to determining the rate of approach. Thus, the bat is endowed with a population of combination neurons that find a match between the frequencies of the call and the echo, and therefore represent and monitor the rate of approach. The algorithm used appears to be another example of computational anatomy; that is, the information is represented as the pattern of activity within a neural map, and changes in the approach velocity are represented as shifts in the position of patterned activity within that map.

Figure 4.15 illustrates one cortical area involved in computing the rate of target approach using Doppler shifts. This region of the mustache bat's auditory cortex is called the CF-CF area because the combination neurons that are found here respond to the constant-frequency component of the call and echo. The CF part of the sonar signal is best suited to measure Doppler shifts because it is comprised of a narrow range of frequencies that are somewhat extended in time. The long duration of the signal is needed in order to get a more precise reading of the frequency.

So CF-CF neurons respond to the combination of the fundamental frequency of the call and one of the harmonics in the echo. Different cells are tuned to the different possible values of the CF fundamental, but an economy of neurons is enabled by the relative stability of the echo. Many of the cells respond to a CF_1-CF_2 combination, where, as before, CF_1 designates the fundamental component of the call and CF_2 designates the second harmonic of the echo.

Suga and colleagues found there is an orderly spatial arrangement of the neurons that populate the CF-CF area. Neurons tuned to high values of CF_1 are found in the back part of the area, and those tuned to the lowest values of CF_1 are found in the front of the area. Like the FM-FM area, this pattern creates one of the coordinates in a cortical map.

While the Doppler compensation system is good, it's apparently not perfect. For one thing, it can't predict the future. Remember that the system has a cycle to it. A call is emitted, the echo comes back, another call, another echo, and so on. So the frequency of the call is determined by the Doppler shift of the last echo. Since both the moth and the bat are flying, the rate of the bat's approach can change very rapidly, which will inevitably produce small fluctuations in the Doppler shifts. Rapid changes in the approach velocity mean that the Doppler compensation cannot eliminate all variation in the frequency of the echo, and so, in order to preserve fine acuity in processing Doppler shifts, combination neurons will have to be differentially tuned to different values of the CF_2 component of the echo. Thus, as we move within the map along any line representing a constant value for CF_1 of the call, the cells are tuned to slightly different values for the CF_2 component of the echo. This is the second axis or coordinate in the map.

This regularity in the arrangement of combination cells means that any particular value of Doppler shift has a corresponding spatial location on the map. Different combinations of CF_1 and CF_2 can represent the same size of Doppler shift, so the pattern of activity can specify the rate of approach. The dark bands in figure 4.15 represent these different rates of approach. The map is arranged such that as the bat approaches with increasing velocity, the pattern of activity within the CF-CF area moves forward: computational anatomy.

We now have some insight into the inner workings of the bat sonar receiver, and some experience with the basic operation of a biosonar system. In the process, we have seen how systems of neurons can compute such things as target range and approach velocity. There remain many other fascinating aspects of this remarkable system. From the neurological point of view, we know that the FM-FM and CF-CF areas of the mustache bat's auditory cortex are only two of eight or more different specialized regions of cortex devoted to processing sonar signals. And that is just the cortex. There are perhaps 10 or more specialized regions below the cortex that are also devoted to sonar processing. Some are involved in comparing the inputs to each ear, which is important for resolving the eccentricity of the target. Some are devoted to producing the feedback to the vocal apparatus needed to achieve Doppler compensation. Some are used to control the amplitude of the call and, as a result, the amplitude of the echo. A lot of work still needs to be done.

Consider, for example, the overall acuity of the Doppler processing system. Behavioral experiments demonstrate that the mustache bat can resolve a shift in the frequency of a 60 kHz echo as small as 6 Hz! This is probably greater than what is needed to compute approach velocity. Is there something else that might be going on? Something else that Doppler shifts might signify to the bat?

Professor Hans-Ulrich Schnitzler, from the University of Tübingen, thought so. He recognized back in 1970 that the actual wing beats from a flying insect could produce fluctuations in the frequency of the echo that could provide information that bats could use to actually identify different species of insects. His first experiments were a model of the elegance of simplicity. He tethered a moth to the end of a narrow metal rod (using glue) and aimed a speaker at the moth. The speaker played long tones at 80 kHz, which simulates the second harmonic of the sonar call of a horseshoe bat. An ultrasonic microphone was used to record the echoes from the moth. The critical comparison was between the echo produced when the moth's wings were stationary and those obtained when the moth was attempting to fly. The results of Schnitzler's experiment are illustrated in figure 4.16.

The echoes recorded when the wings were stationary are illustrated graphically on the left. Each horizontal trace represents a small sample

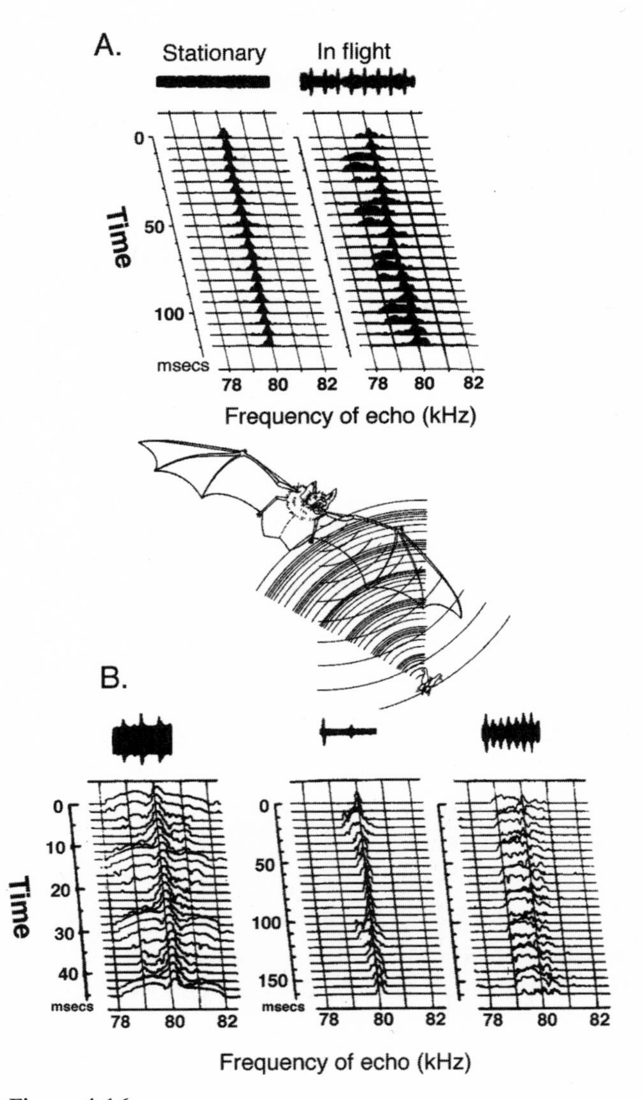

Figure 4.16
Recordings of the echoes of an 80 kHz tone produced by different species of moths. The wingbeats produce changes in the component frequencies of the echo. These changes are called glints and are illustrated in A. The echo produced when the wings are stationary are narrowly restricted around 80 kHz, whereas the range of frequencies in the echo is much broader when the wings are beating (labeled "in flight"). Panel B illustrates how the glints differ for three different varieties of moth. The time domain representation of the echoes are illustrated in the upper part of each panel. The lower part of each panel are spectrograms of the echoes. The spectrograms show more clearly how the frequency components vary over time.

of time, and the series of traces indicates how the echo changes over time. The bottom line? It doesn't. The little black regions remain centered at 80 kHz, indicating that the echo is a constant 80 kHz signal. Look at the traces on the right, however. These traces were obtained when the moth was beating its wings. The echo is obviously different. It contains more frequency components (as indicated by the increased width of the blackened regions), and (as indicated by the changes in the blackened regions), the frequency components change over time. In fact, you can actually see the wing beats in these records. So this experiment shows that the information is there, it's available. The question is: Does the bat use it to actively select which insects it wishes to catch?

5

Variations on a Theme: Sonar Beneath the Seas

If it's a good idea, use it. Again and again. That's certainly what nature does. Of course, when a good idea is applied in a new situation, some adaptations have to be made. The principle may be the same, but the actual implementation may have to be modified. A little creativity may be required. Thus it is with biosonar.

Bats aren't the only creatures that have evolved this sixth sense. For some time it's been known that certain cave-dwelling birds have it. Clearly, the advantage of biosonar is greatest when vision is not possible. When there is light, animals evolve a sense of vision. When there is no light, they become creative. The oceans represent another area in which light quickly becomes a precious commodity. The amount of sunlight penetrating the oceans decreases dramatically with each additional meter in depth. In the deeper regions of the oceans, it is always nighttime.

So, for aquatic animals, there was certainly a reason to evolve a biosonar system. There was also opportunity: sound waves are transmitted in water very efficiently. Of course, not every creature is capable of evolving a sophisticated sonar system. We've learned from the bats that it requires a pretty fancy nervous system, something most marine invertebrates don't have. Likewise with most fish.

There is one class of marine animal that is endowed with a very elaborate nervous system, however: the marine mammals. The result? Most, if not all, of the marine mammals have developed biosonar capacities. Soon after Griffin and his collaborators established the sonar abilities of bats, people began to investigate the possibility that dolphins had a sonar system of their own. In the late 1940s Arthur McBride, who was the first curator of Marine Studios in Florida (which later became Marineland) speculated about the possibility of dolphin sonar in his personal notes (published posthumously in 1956). The observations and experiments that established the use of sonar in dolphins paralleled those that had proved so successful in bats. It's a familiar routine now. First, establish that the dolphins can navigate or discriminate different objects without using vision (or touch). Then, see if the dolphins emit sonarlike signals when they are exploring their environment or performing obstacle avoidance or object identification tasks. If they produce ultrasonic emissions when performing these tasks, then try to establish that they can actually hear the echoes.

McBride made observations of the first kind: he noted that the Atlantic bottlenose dolphins could avoid fine nets used to capture them. Since these dolphin-catching expeditions often occurred at night in brackish estuaries, visibility was quite poor. Despite this, McBride noted that the dolphins would sometime swim over a low-lying net. They appeared to distinguish details of their surroundings in settings that made visual guidance unlikely. It reminded McBride of the bats. Unfortunately, McBride died in 1950, and his notes were not discovered until after his death. In the early 1950s, other marine biologists independently began to suspect biosonar in dolphins; the first paper to explicitly suggest it appears to have been written in 1952 by Kellogg and Kohler. They reported the results of learning experiments that established that dolphins could hear sound frequencies as high as 50 kHz. Later work indicated that dolphins could discriminate objects of different sizes and different materials without visual aid (fortunately, Spallanzani's methods were not needed—dolphins can be coaxed into wearing opaque occluders over their eyes; see figure 5.1). The third essential observation, that the dolphins emit ultrasonic calls during navigation and object identification tasks, was made in 1952 by Forrest Wood, who succeeded Arthur McBride as the curator of Marineland in Florida. These early observations provided the first real evidence of biosonar in dolphins.

Of course, dolphins are fascinating creatures to humans. The size of the dolphin brain is entirely comparable with the size of a human brain.

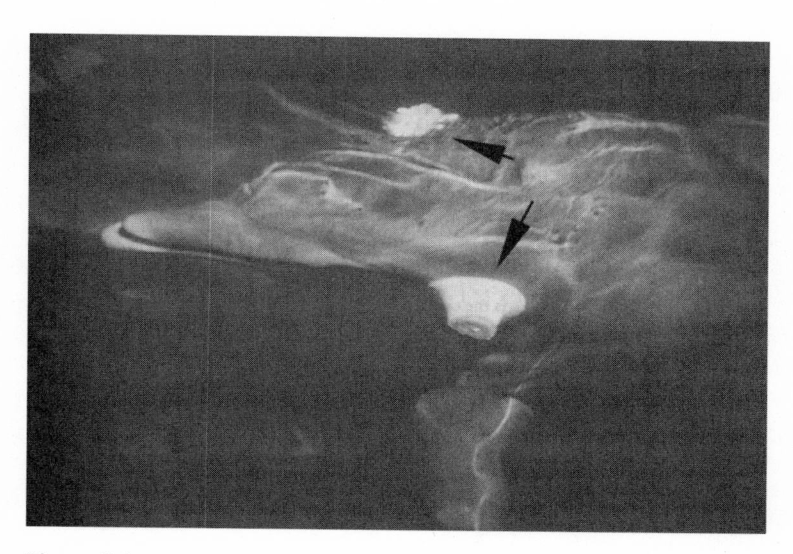

Figure 5.1
A dolphin wearing suction-cup opaque eye covers used to test for sonar abilities in the absence of vision.

Figure 5.2 compares the brain of a dolphin with the human brain. I think you'll agree that the resemblance is striking. Dolphins are widely recognized for their intelligence. They also reportedly have an altruistic bent: they sometimes save swimmers and mariners in distress. Like bats, dolphins are a part of our folklore. They get substantially better press than bats, however. Dolphins are cute and entertaining. Like bats, dolphins are incredibly athletic. All the world adores athletes, but while people flock to see acrobatic dolphins in marine parks, there is apparently little interest in paying admission to see a bat show.

The dolphins aren't the only sea creatures that possess biosonar. Many of the marine mammals apparently possess sonar abilities. Certainly all those that have been studied emit ultrasonic calls, which is at least prima facie evidence of biosonar. The list includes beluga whales and killer whales. The operation of the biosonar systems of marine mammals is not nearly so well understood as that possessed by bats. There are many reasons for this. It is more difficult and much more expensive to study large marine mammals than small flying ones. It is also not possible to perform many types of experiments on dolphins. Scientists have learned

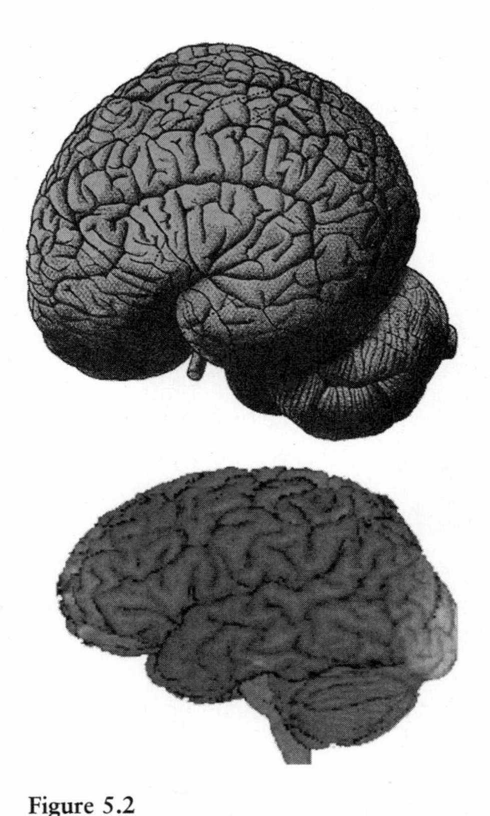

Figure 5.2
A comparison of the dolphin (top) and human (bottom) brains. The dolphin brain was drawn by the noted comparative neuroanatomist G. Elliot Smith in 1902. It was reproduced in Ashley Montagu and John C. Lilly, *The Dolphin History,* published in 1962. The human brain is a magnetic resonance image of a patient at the Epilepsy Unit of the Dartmouth–Hitchcock Medical Center. The skin and skull are removed via computer-software. The computer rendering was done by Dr. Terrance M. Darcey, Department of Neurology, Dartmouth–Hitchcock Medical Center. The similarity between the brains of these two species is quite remarkable. Although they are slightly different in shape, the cortex of each is very well developed. Note the convolutions (wrinkles) in the cerebral cortex. This is characteristic of intellegent animals and, if anything, is more conspicuous in the dolphin than in the human brain.

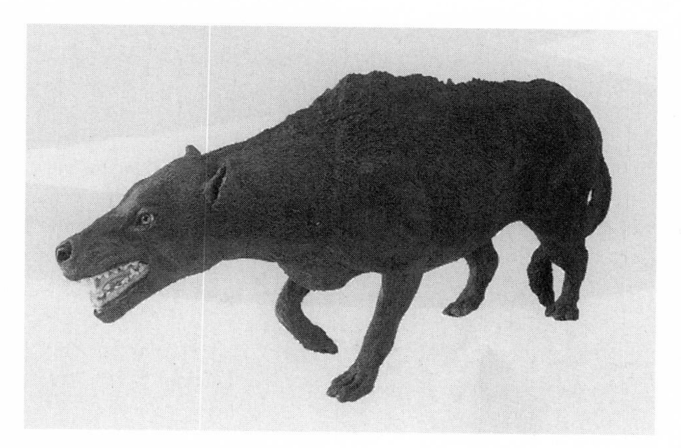

Figure 5.3
Paleontolologists believe that dolphins evolved from this unusual creature called a mesonychid, which roamed riverbanks some 55 million years ago.

some remarkable things about the biosonar system of marine mammals in general, and dolphins in particular. It can only add to our admiration of these wonderful creatures to learn a little more about their ultrasonic view of the world.

As an interesting aside, I'll mention that the earliest dolphins appeared in the oceans at about the same time the first bats appeared on land, about 50 million years ago. So, like the bats, dolphins have had some time to perfect their biosonar systems. Paleontologists believe that dolphins evolved from a primitive mammalian carnivore called a mesonychid, a strange beast that had hoofs but otherwise looked like a wolf (figure 5.3). This unusual creature is suspected of gradually adapting, over many generations, to a lifestyle that may have been something like that of a hippotamus—spending more of its time in water than on land. Gradually the forelegs evolved into flippers, the hind legs became vestigial, and the tail developed into a powerful fluke that propelled the early dolphins through the water at great speeds. Modern dolphins have been around for only about 5 million years—roughly as long as modern humans (Heyning, 1995). They represent another marvelous example of the astonishing adaptability of life on Earth.

6

A Different Kind of Sonar Transmitter: The Dolphin Call

As before, we should start at the beginning, with the call. In addition to its advantage as a logical starting place, the call has been studied extensively, using techniques that range from inserting tubes down the blowhole to magnetic resonance imaging of dolphin heads during the production of calls. So at least that much is decided. We'll start with the call.

Perhaps the most interesting thing about the call is that it took researchers a long time to figure out which part of the dolphin produced it. One prevalent view held that the call came from the larynx. Just like the bat. Just like us. Indeed, just like the audible calls that dolphins can make when they are above the surface, and want to "talk" to us.

The data to support a laryngeal source were not that strong. First, there was an anatomical argument. The larynx of the dolphin is complicated. It looks as complicated as that found in humans. Why, the argument goes, would evolution provide for such a complicated structure if not to produce vocalizations? A simple argument based on anatomical complexity. We've seen our intuitions, our common sense, lead us astray before. Some caution seems warranted here.

The laryngeal theory was not based entirely on anatomy, however. There was some additional evidence suggesting the larynx might be the source of sonar signals. You can decide for yourself how compelling you think it is.

These experiments were performed on the severed heads of some dolphins. I don't know where the heads came from. Maybe they came from dead dolphins that had been caught in fishing nets. That was before the laws in the United States required "dolphin safe" tuna fishing technologies. Now every can of tuna sold in the United States advertises its virtue by assuring the consumer that no dolphins perished to provide them with their tuna salad.

What was done with these heads, however, once they were obtained? Well, compressed air was injected into the larynx from behind, in an attempt to see what kinds of noises could be produced. Noises were indeed produced. They are often described as "barks" or "whistles." Under certain conditions, high-pitched tones could be produced. But these artificially produced sounds weren't very good facsimiles of a dolphin sonar call. Well, proponents of the laryngeal theory might say, "What can you expect? They aren't live dolphins after all, just dolphin heads. The conditions are clearly not ideal. A live dolphin would have control over the musculature, could shape the call. This simply showed that the larynx could in principle be a source of vocalizations." Not a very satisfactory state of affairs.

Then came a series of experiments that attempted to identify the anatomical source of the calls. These experiments used a wide variety of technologies, and were often quite creative. Early workers placed an array of hydrophones (underwater microphones) around a cooperative dolphin's head while it emitted sonar signals. Using detailed information concerning the times at which each hydrophone recorded the call and some mathematical calculations, this work localized the sonar source in the vicinity of a part of the nasal cavity called the nasal plug. (See figure 6.1.)

So it looked as if dolphins might generate sonar calls by using their nasal system rather than their larynx. Other work seemed to confirm this. High-speed X-ray motion pictures and soft-tissue spot X-rays of several species of dolphins producing sonar signals showed movements within the nasal system but not within the larynx. Even the Doppler effect came into play. The Doppler effect can be used to measure very small movements of a surface by producing an ultrasonic sound wave that is directed at the surface. Small movements of the surface produce Doppler shifts of the echo that specify exactly how much the surface is moving. These ultrasonic Doppler motion detectors can pick up movements of less than

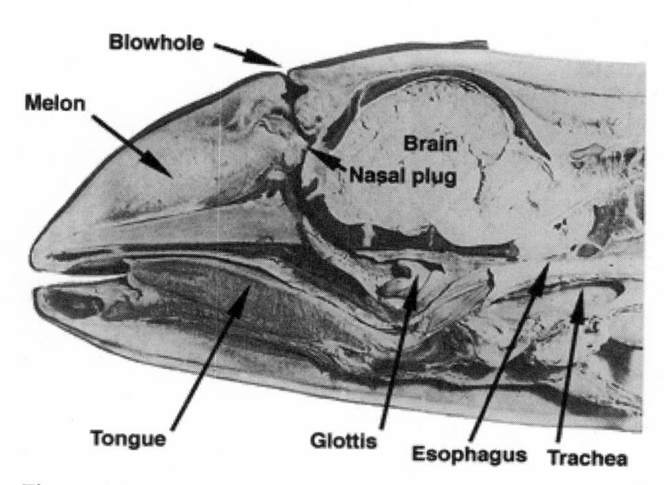

Figure 6.1
A view of the midline of a dolphin's head (the head has been cut in half lengthwise). The positions of the brain, the blowhole, and the larynx (bounded by the glottis at one end and the trachea at the other end) are shown, as is the nasal plug. The role of the football-shaped melon will soon become clear.

10 millionths of a meter! Using this device, it was again observed that the nasal system vibrates in synchrony with the sound, while the larynx is quite still.

I mentioned earlier that the methods used included inserting a tube into the blowhole. To do this, you obviously need a cooperative, trusting dolphin. Once the tube is inserted, the dolphin closes the blowhole around it. If the tube is sealed, no air escapes when the dolphin tries to increase the nasal pressure. In this case, the animals emit sonar signals without difficulty. If however, the tube is left open, the dolphin cannot pressurize the nasal system. Under these conditions, the animals are unable to produce ultrasonic calls. When considered together, the evidence for a nasal source is pretty impressive.

Where does the sound go after that? How is the sonar signal propagated through the dolphin's head? How does it get into the water? It's not as obvious as it might seem. To see why, we have to review some facts about acoustic impedance.

We first encountered issues associated with acoustic impedance in considering the function of the little bones in the middle ear, the ossicles.

Recall that when a sound wave encounters an interface between two media, some of the energy is reflected from the interface, and some is transmitted into the second medium. If the impedance of the two media is very different, then most of the energy is reflected and very little is transmitted into the second medium. This is the case at an air–water interface. The acoustical impedance of air is much less than that of water, so the vast majority of acoustic energy traveling in air is reflected when it encounters an air–water interface: very little energy is tranmitted into the water. The function of the middle ear bones is to overcome the impedance mismatch that occurs as airborne sound waves try to enter the fluid-filled cochlea.

The dolphin faces what is essentially the inverse problem: If sound waves are created in the air-filled nasal cavity, how will they get transmitted into the seawater? There is an air–water interface, and so there is an impedance mismatch. If the dolphin cannot overcome this impedance mismatch, the sonar signal will not get transmitted into the seawater; it will simply bounce around in the dolphin's head. Not a desirable state of affairs.

This problem of impedance matching served to focus attention on an interesting structure that is unique to dolphins. The structure is called the melon. It is a large saclike affair made of fat tissue. The melon lies beneath the dolphin's very prominent forehead. The tissue of the melon has some interesting acoustical properties. First, it transmits sound energy quite well. Not only that, the melon is made of different types of fat. One type is very peculiar, and is found only in the melon and around the dolphin's lower jaw.

One of the main features of an acoustical transmitting medium is the speed at which sound waves travel through it. Air, with its low impedance, transmits sound waves at a relatively low velocity. In contrast, water, with its high impedance, transmits sound energy at a much higher velocity (344 meters/sec for air, 1376 meters/sec for water). One way to overcome an impedance mismatch is to do what the middle ear bones do: convert a small force distributed over the large surface of the eardrum to a large force distributed over the much smaller oval window. Another is to change the velocity of the sound wave precisely so that it matches the sound velocity characteristic of the second medium. So, if the velocity

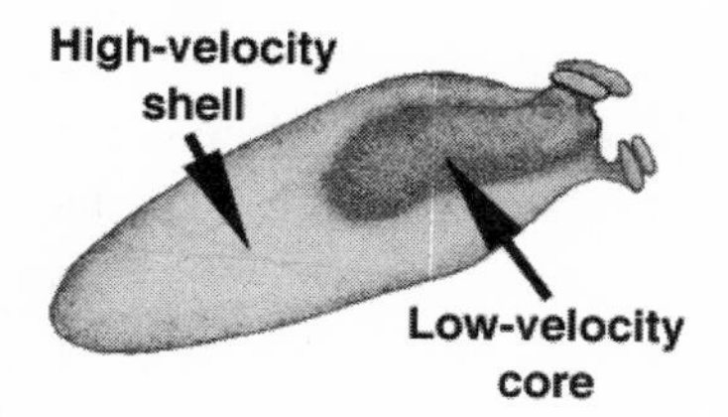

Figure 6.2
The melon, which lies in the dolphin's forehead, serves as an impedance matching device that improves the efficiency of sound transmission from the system of nasal sacs to the water. View is as if dolphin were facing left.

of sound could be increased within the melon, the melon could serve as an impedance matching device: something that would efficiently transmit the sonar signal from its origin within the nasal cavity through the dolphin's head and into the water.

The test of this idea was conceptually simple. One must measure the velocity of sound as it travels within the melon. Such measurements were made in 1974 by K. S. Norris and G. W. Harvey. They obtained the melon from a Pacific bottlenose dolphin that had just died. Norris and Harvey then cut the melon into thin slices and measured the time it took for a sound wave to traverse the thickness of each slice. This was done by placing sensitive hydrophones on both sides of each slice, and measuring the time at which a sound wave entered one side and exited the other. Given the thickness of the section and the time interval between the response measured by the first and second hydrophones, it is a simple matter to compute the sound velocity through the slice. Norris and Harvey found that the velocity of sound transmission is different in different parts of the melon. In general, there is a low-velocity core surrounded by a high-velocity shell. Subsequent work using three-dimensional reconstructions of CT scans of the melon reveal a low-density core surrounded by a higher-density shell. These reconstructions of the melon are illustrated in figure 6.2. The two zones correspond well to the velocity profiles described by Norris and Harvey.

Notice the little appendages at the back end (on the right). They are called bursae, and are attached to the nasal cavity within the dolphin's head. Since the nasal cavity appears to be the source of the sonar signals, the acoustic energy could be transferred from the nasal cavity to the bursae. Then, as the sound waves travel forward within the melon, their velocity would increase. As they approach the front of the melon, the sound waves would be traveling at about the same speed as they do in seawater. Then the signals could be efficiently transmitted into the water, and the reflection would be minimized. The melon appears to serve as an impedance matching device: it permits the dolphin to generate sounds in an air-filled nasal cavity and effectively transmit that energy into the water. What an elegant solution! In fact, man-made sonar transducers work similarly. They use a mineral oil interface between the crystals that produce (or detect) sound vibrations and the seawater.

These variations in sound velocity within the melon may serve an additional purpose: that of focusing the sonar beam. Whenever the velocity of an acoustic or electromagnetic wave changes, the direction of propagation changes as well. This change in direction is called refraction. We are all familiar with the ability of a convex lens to refract or bend light rays—it is the basis for all optical imaging systems, such as a telescope, a microscope, a movie projector, a camera, or an eye. Similar effects can occur with sound waves. In fact, sound waves follow the same laws of reflection and refraction as light waves do. If this were not the case, we couldn't speak of sonar systems creating an acoustic image of the environment. In any case, the increased velocity of sound waves as they are transmitted through the melon also refracts or bends the sound waves, serving to focus the sonar signal into a beam that is directed in front of the dolphin (figure 6.3).

The shape of the sonar beam of dolphins has been very carefully described by W. Au and coworkers at the U.S. Naval Ocean Systems Center in Hawaii. Dr. Au has shown that in every marine mammal studied to date, sonar signals are emitted from the forehead as a focused beam of acoustic energy. The size of the beam varies to some extent in different species, but it is always directed straight ahead of the animal, and is usually elevated in the vertical plane by some 5 or 10 degrees. We saw earlier that the sonar beam of bats is also somewhat focused, and tenatively attributed at least part of the focal power to the elaborate nose-leaves found on many echolocating bats. It appears that the melon serves a similar purpose for the porpoise. The melon is therefore sometimes called an acoustic lens, since it has the ability to focus the sonar

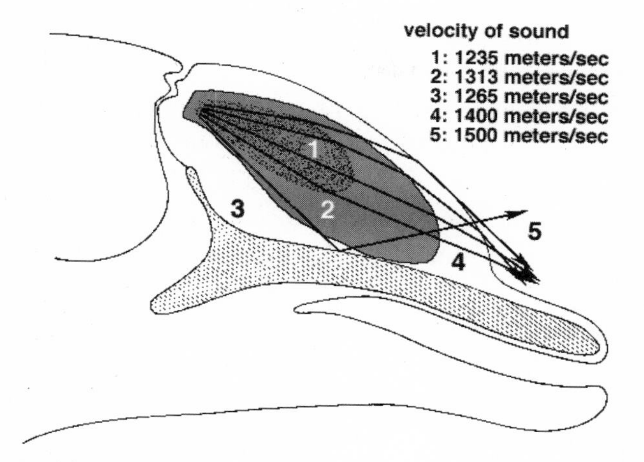

Figure 6.3
The melon acts like an acoustic lens that focuses the sonar beam in front of the animal.

beam in a manner that is analogous to an optical lens. In the interests of completeness, however, we should note that numerical calculations performed by Dr. Au suggest that the sonar beam of a dolphin is more narrowly focused than might be expected on the basis of the refractive power of the melon alone. The bony structure of the dolphin's head may also make an important contribution in this regard.

Now we have some appreciation for the elaborate biological engineering that permits the dolphin to generate sonar signals within its nasal cavities, and efficiently transmit those signals into the surrounding water. Like those of bats, dolphins' sonar signals are highly directional, forming a beam of acoustic energy that "illuminates" the region immediately in front of the animal. Also as with bats, the sonar signals of marine mammals are very intense.

We know why. The more intense the signal, the longer the effective range of the sonar system as a whole. The effective range of marine biosonar signals is difficult to evaluate in any absolute sense, since the range at which an object can be detected depends upon the size of the object. However, what we do know is pretty impressive: dolphins can detect the presence of a water-filled steel ball 7.5 cm in diameter at distances up to 120 meters. In order to put that observation in perspective, if you were to *look* at that same steel ball from a distance of 120 meters, the image on your retina would be very small indeed. The diameter of the ball's image as projected on your retina would be about 0.010 mm or 10 micrometers (μm)! This corresponds to the area covered by about four of the smallest photoreceptors in the human eye: those found in the most acute part of the retina, called the fovea. Whether you could detect such a small image would depend on the background against which the ball is viewed. If the background was very different from the ball in brightness or in color, you could see it. But the conditions would have to be favorable. The same is true of marine biosonar.

For instance, dolphins often swim in shallow water, and their sonar signals are reflected off the sea bottom and off the surface of the water (where, once again, we have an air–water interface). Reflections from the bottom represent a major factor limiting the range of sonar, because the reflections from these geological formations can interfere with or mask the much weaker echoes from smaller objects of interest, such as fish, other dolphins, or those curious steel balls used by human experimenters. Experiments done in the vicinity of a major ridge along the bottom of the Kaneohe Bay yielded an estimated range of only 77 meters, as opposed to the 120 meters where the bottom is smoother. Since the two experiments used the same dolphin and even the same type of target, it would appear that this substantial disparity is attributable to the differences in the bottom in these different regions of the bay. While the surface reflections are always present, they do not appear to be a major source of interference for the dolphins.

Exactly how intense are dolphin sonar signals? As impressive as the intensities of the bat calls were, the dolphin calls are even more energetic. The first recordings of marine sonar calls were made in large aquarium tanks. Recorded intensities under these conditions were about 170 dB. It turns out, however, that dolphins can produce sonar calls far in excess of 170 dB. This was first shown by Whitlow Au and colleagues, who were the first to record marine sonar signals in the open water. In Hawaii's Kaneohe Bay, dolphin calls reach intensities as high as 220 dB. This means that the sonar calls of dolphins in Kaneohe Bay were about 500 times more intense that those observed in aquarium tanks! In the tanks, the dolphin's sonar signal is a whisper in comparison with what it produces in the open ocean. What accounts for the difference?

It appears that there are two contributing factors. First, like the seafloor, the sides of the aquarium tank are excellent reflectors of sound energy. As as a result, sonar signals reverberate off the sides and bottom of the tank, and the echoes are very intense. To the dolphin, it might seem a little like sitting inside a big bass drum while the band plays a rousing Sousa march. The dolphin reduces the amplitude of the call in order to maintain a reasonable amplitude on the echo. Consistent with this suggestion, it has been shown that dolphins will indeed produce calls in excess of 200 dB if the pool is lined with a sound-absorbing material (water-soaked cypress panels).

The second important factor that contributes to the amplitude of the signal is the ambient noise level. Au and colleagues recorded the sonar calls of a beluga whale in San Diego Bay, then moved the whale to Kaneohe Bay in Hawaii, and recorded the whale's sonar signals there. The signal intensities were 202 dB in San Diego, and increased to 210 dB in Kaneohe Bay. This represents about a three-fold increase in sound energy. Interestingly, the noise level in Kaneohe Bay is about 15 dB (fifty times) higher than in San Diego Bay. One major source of this noise is the large population of snapping shrimp in Kaneohe Bay. It turns out that, in addition to its tropical beauty, Kaneohe Bay has the highest level of snapping shrimp noise in the world! The noise level in Kaneohe Bay might be a factor in the sonar range experiments that were just described: a quieter bay could easily reveal a detection range even more impressive than the current estimate of 120 meters.

These sorts of observation imply that dolphins and other marine mammals exercise a considerable degree of control over their sonar signals. They adaptively adjust the intensity of their sonar calls to maximize performance. If the environment has a lot of reverbation, they lower the intensity. If there is a lot of ambient noise (like that from those pesky little snapping shrimp), they increase it. It is frequently found that the dominant frequency of the dolphin call increases as the intensity increases. It appears that this increase in frequency is secondary to the increase in intensity: that is, dolphins and other marine animals must increase the pitch of their calls when they make them louder. It has been suggested that they cannot make a lower-pitch call very loudly, but perhaps they can make a high-pitched call more softly. Control over the dominant frequency could be an important means of improving sonar performance in the presence of noise, but only so long as the noise is composed of a relatively narrow range of frequencies. When the noise is confined to a narrow frequency range, producing a call whose frequency characteristics are different from the noise will improve sonar performance, if the sonar receiver can "tune in" to the signal frequency and "tune out" or filter the noise frequency. This is a standard way to increase sensitivity in a variety of signal-processing applications. Unfortunately for the dolphins, measured noise levels typically show that the noise contains a great deal of energy distributed over a wide range of frequencies (we term this "broadband noise," to emphasize the wide range of frequencies it contains). Thus, a great deal of frequency control in marine biosonar may not be very advantageous in overcoming noise within the sea. Perhaps this is why the dolphins do not appear to have developed fine-tuned control over their sonar frequencies. To evolve such control might require elaborate changes in the sonar transmitter, which, as we have seen, is already quite complicated. Evolution, in its constant search for efficiency, may have decided that volume control is enough, and frequency control is simply not worth the effort.

It turns out that marine mammals make a number of different sounds, and they are not all sonar signals. Some appear to serve as a means of communication. Although there is much in folklore and the movie industry to suggest that dolphins and other marine mammals attempt communication with humans, these vocalizations almost surely evolved to communciate with other members of the same species. These mammals often have elaborate social systems, so the need for communication is almost surely strong. Many of the social signals are in the audible range of humans, and it appears that these may indeed be produced in the larynx (for example, it is known that dolphins can produce audible sounds and ultrasonic sounds simultaneously or separately, which clearly implicates at least two independent sound generators). Here we are most interested in the ultrasonic sonar calls, and they have some important differences from those that characterize bat calls.

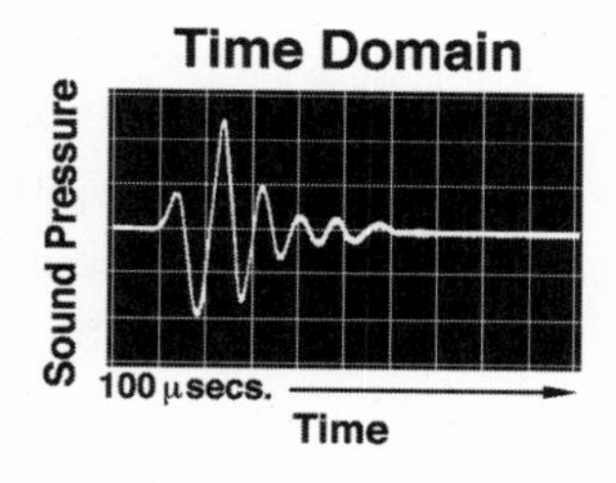

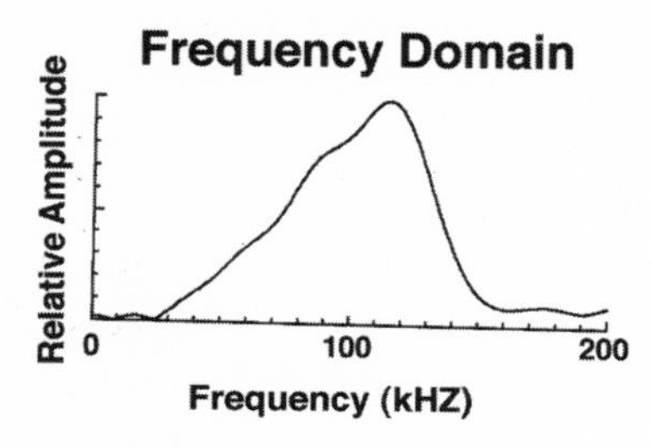

Figure 6.4
The sonar signal emitted by a bottlenose dolphin. The top panel shows the waveform as it might be recorded using a hydrophone (the figure actually shows the average of several sonar "clicks"). The bottom panel shows the relative amplitude of the different frequencies that make up the signal, which is called the amplitude spectrum. Graphs called spectrograms were used to display the important features of bat sonar calls, because the frequency characteristics of bat calls vary over time. Spectrograms illustrate how the amplitude spectrum of a signal varies over time. The dolphin sonar signals are so brief that we don't need to worry about time so much. As a result, the amplitude spectrum is sufficient, and we don't have to use spectrograms to study dolphin sonar.

The waveform of a typical dolphin call is illustrated in figure 6.4. It looks like a sine wave that grows and decays very rapidly. It fact, that is very close to what it is. Notice that there is no temporally extended constant-frequency component (like that seen in CF-FM bats), and there is no downward sweep in frequency (the FM component in the CF-FM bats and the pure FM bats). These calls are called clicks, but they are really more like little tone bursts. In the lower panel of the figure, we see the signal graphed in a different format. Rather than showing how the pressure varies with time, as in the panel above, it shows the different sine wave frequencies that, when added together in the correct manner, produce the waveform in the upper panel. These representations are completely complementary and equivalent. Using the right mathematics, one can go from one form or representation to the other. The upper panel is what is seen when recording a sonar call with a hydrophone. As with bat calls, you could hear a rendition of the call by replaying the recording at a slower speed. It would sound like a click because it is so very brief (after all, as the time scale indicates, the call lasts less than 100 millionths of a second (μsecs), so even if played at one-fifth the normal speed, the call would be over in much less than the blink of an eye!). The lower panel is different, in the sense that it tells us the different frequency components in the call. We see that the dominant frequency (the one that is most energetic) is about 110 kHz (110,000 cycles/sec). However, the figure also tells us that substantial energy exists in a frequency range extending from about 30 kHz to 150 kHz. Graphs such as the one in the lower panel are called frequency spectra, a term meant to emphasize the analogy between adding up these frequencies and adding up different colors of spectral light to get any arbitrary hue).

One thing the frequency spectrum makes clear is that dolphin sonar calls are quite different from the sonar calls of CF-FM bats. In CF-FM bats, the energy is confined to a fundamental frequency and several harmonics (integer multiples) of that fundamental. In dolphins, we find the call has many more frequency components. Dolphin sonar signals are therefore called broadband calls. Often, people refer to waveforms like that in the upper panel as *time domain representations,* and graphs like the one in the lower panel as *frequency domain representations.* While it is important to recognize that each representation can help us see different aspects of the signal, it is equally important to remember that they are equivalent: mathematics allow us to go from one to the other.

Why do we need to consider two different representations? Well, to some extent, we've already seen how a graph of the frequency spectra clearly indicates important differences between bat and dolphin sonar calls. But there is a deeper reason. We might say that the production of the signal, the sonar transmitter, works in the time domain: little membranes are tensed and air passing by them causes vibrations that in turn produce sounds. On the other hand, much of what we learned about bat sonar suggests that *processing* of the echo by the sonar receiver operates in the frequency domain—it analyzes the different frequency components separately, which is a neural analogue of the frequency domain representation shown in the lower panel of figure 6.4. This analysis in the frequency domain began with the mechanical properties of the basilar membrane in the cochlea. Mechanical features such as the width, thickness, and compliance of the membrane resulted in a membrane that vibrates at different places according to the frequency of the impinging sound waves. Then these local motions of the membrane were sensed by hair cells, which in turn activated nerve fibers. Different frequencies are, in effect, spatially segregated along the length of the basilar membrane. As a result, the hair cells respond in a frequency-selective manner, and because each fiber in the auditory nerve contacts a limited number of adjacent hair cells, these fibers also respond selectively to a narrow range of frequencies. The pattern of activity across a set of frequency-specific fibers represents a biological implementation of a frequency domain type of analysis, so it is good to have some understanding of the frequency domain style of thought as well as the more intuitive time domain representations of acoustic waveforms.

Here is a little more. A sort of "cosmic truth" about dolphin sonar signals. There is an interesting relationship between the duration of a signal and the frequency components it must contain. If the signal is very long lasting, it *may* contain a very narrow range of frequencies. Consider, for example, the concept of a sine wave. It is a mathematical construction that is never fully realized in nature. We think of the sound wave produced when we strike a tuning fork as a "pure tone" (a sine wave), but technically that is not quite correct. The reason is that when we strike a tuning fork, the sound begins and, after a certain amount of time, it ends. This everday fact has an important consequence for the frequency domain representation of the sound produced by striking a tuning fork.

If we were to strike a tuning fork and record the sound it makes, we would see the waveform grow, look sinusoidal for a period of time, and then diminish. And, if we were to obtain the frequency spectrum from that waveform, we would see that it, like the dolphin sonar call, is composed of a *range of frequencies*. In the limit, a true sine wave *must* be infinitely long: only then will a spectral analysis show that it is "spectrally pure." And the purity of a sine wave is revealed by its frequency spectrum: a spectrally pure sine wave will have only one spike, one bold vertical line signifying to all the world that this signal contains only one frequency.

What's more, the shorter the signal, the more frequencies it must contain. In the limit, an infinitely brief signal contains all frequency components. So there is a principle of mutual exclusion here. If the signal is to be spectrally pure, it must be long lasting. This obviously will not work for the dolphin, however, for the dolphin must emit a sonar pulse and await the returning echo. If it were to make a long-duration call, the echo and the call would be mixed in ways that would complicate sonar processing. Temporally discrete calls are needed to compute call–echo delays, which specify target range. The sonar system relies on temporal comparisons between the call and the echo, and since sound travels so much faster in water than in air, short-duration calls are absolutely essential. It would, however, still be desirable to have the energy at least centered on as narrow a range of frequencies as possible, so that the nervous system does not have to distribute all the neural resources needed for biosonar throughout the entire range of frequencies.

What we have is a fundamental incompatibility between resolution in both the time and the frequency domains: if the signal is long, we gain spectral purity but lose temporal information. However, if the signal is brief, what we gain in temporal resolution is lost in frequency resolution. The problem is in many ways analogous to the famous uncertainty principle of Walter Heisenberg: one can know the frequency with arbitrary resolution, but only with a consequent loss in temporal sensitivity. What we need is a solution. Not something that gets around the uncertainty between frequency and time; this cannot be done. But what can be done is to find an optimal compromise, a waveform that maximizes both the temporal information and the frequency information.

Is there a principled way to arrive at such a compromise, or would one have to resort to trial and error? This problem was addressed by Dennis Gabor, a Hungarian-born physicist who emigrated to Great Britain in 1933. Dr. Gabor became a naturalized British citizen and, after working for a number of British companies, joined the faculty of the Imperial College of Science and Technology at the University of London in 1949. Gabor invented the system of three-dimensional photography called holography in 1947, and for this achievement was awarded the

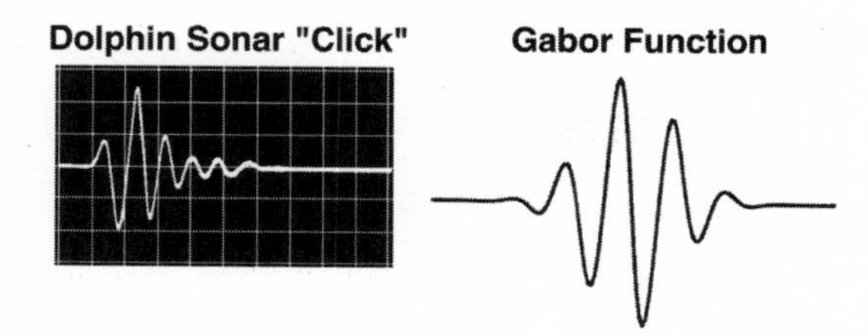

Figure 6.5
The striking similarity between a dolphin sonar click and the Gabor function, which represents the ideal compromise between short duration and spectral purity.

Nobel Prize in physics in 1971. In 1967 he was appointed a staff scientist of the Columbia Broadcasting System Laboratories in the United States, where he worked on the uncertainty principle as it relates to the mutual incompatibility of temporal resolution and frequency resolution of waveforms. It turns out that this was an important problem in the communications industry. Dr. Gabor discovered that the detector that optimizes resolution in the time and frequency domains has a specific mathematical form. Filters that implement this optimal compromise are now called Gabor filters. An example of a sinewave passed through a Gabor filter is shown in figure 6.5. It looks alot like a dolphin sonar signal, doesn't it? In fact, Gabor waveforms provide an excellent mathematical model of a dolphin sonar call. The Gabor waveform represents a mathematical description of an optimal compromise. If you want to produce a signal that is maximally restricted in both time and frequency, you can't do any better. The beauty of it is that the dolphins seem to have come upon the same solution: dolphin sonar signals approach a mathematically idealized sonar signal. It is cause for a little reflection. I don't know which is more inspiring: that the human mind can, through mathematics, solve such subtle problems, or that the dolphin, through evolution, can do the same.

How do dolphins use their sonar system? We might guess that, as with the bats, dolphin sonar is used to navigate and avoid obstacles, to find and capture prey, and to detect and avoid predators. Those are pretty much the main functions of any sensory system: in short, to inform the perceiver about what is "out there." There are apparently few studies of how dolphins or other marine mammals use their sonar abilities in their everyday lives. Most of the information we have about the manner in which sonar signals are actually used by dolphins comes from experiments in which the dolphins are trained to indicate whether or not they can detect an object by using sonar alone, or whether they can discriminate between the sizes or shapes of different objects (again, relying entirely on sonar). We touched on this type of experiment when we considered the range of the sonar system.

One of the salient aspects of sonar emissions is their repetition rate. When dolphins are searching for a target, they emit a series of those sonar clicks that so closely resemble Gabor functions. There are some interesting aspects concerning the intervals between these clicks. Like bats, dolphins typically emit a click and wait to receive the echo before emitting the next click. As a result, the time interval between the call and the returning echo is a function of the distance from the target: the longer the distance, the longer the interval between successive clicks. If the dolphin emits each click a fixed amount of time after the arrival of the each echo, we would expect the intervals between successive clicks to decrease in direct proportion to decreasing distance to the target. This is in fact what has been consistently observed, at least with respect to the average click rate. We noted this same behavior as bats approached a target, and referred to it as the "feeding buzz."

The close relationship between target range and click rate clearly indicates that dolphins have fairly precise control over the timing of their sonar emissions, although the control is not perfect. There is some variation between successive click intervals even when the target range remains constant. One might surmise that some processing of the echo takes place during the interval between successive clicks, and one likely source of variability in the interclick interval is variablity in the time the dolphin takes to process each echo.

One interesting aspect of dolphins' control over click intervals is that it appears to be heavily influenced by their *expectations*. For example, in a target detection experiment, the target is physically present on some trials, and on others it isn't. Usually, half the trials have a target and the other half do not. Of course, the dolphin's task is to tell us (by making some relatively arbitrary learned response selected by the human experimenter) whether or not it *thinks* a target is present. If the responses are consistently correct, then we can infer that the dolphin can tell whether or not the target is there. As the task is made increasingly difficult, the dophin's performance deteriorates. When accuracy reaches the chance rate, we can infer that the dolphin can no longer detect the target.

In many experiments, target distance is kept constant over a series of trials, so for the target-present trials, the click–echo delay is constant. Naturally, there is no target echo on target-absent trials. This raises an interesting question: How long are the intervals between successive clicks on target-absent trials? The answer is that they are only slightly longer than those produced on target-present trials. What this implies is that the dolphin has a good sense of *when the echo should arrive*. If no echo is detected, then the dolphin waits only a few additional milliseconds beyond the expected arrival time before emitting the next click. This clearly suggests that the dolphin has an accurate sense for the interval between click and returning echo when the target is present.

What happens in a target detection experiment when the target range varies from one trial to the next? Under these conditions, the click intervals on target-absent trials are only slightly longer than the click-echo delay associated with the longest target range in the sequence, suggesting that the animal can maintain an accurate idea of each of the target ranges possible in a given experimental setting. Not only do these observations imply that the dolpbin has an accurate representation of different click-echo delays, but they also require an accurate representation of a number of such delays in their memory.

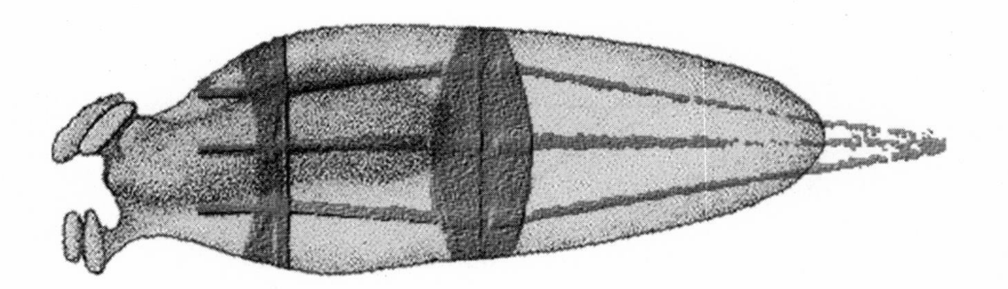

7

The Dolphin's Sonar Receiver

As in the bat, the auditory system is the dolphin's sonar receiver, and dolphins also possess a middle and an inner ear. Unlike terrestrial animals (and certainly unlike the microchiropteran bats), the dolphin does not posses an outer ear (the technical term for the external part of the ear is the *pinna*). The reason for this lack of a pinna is hydrodynamic: in order to swim underwater at high velocities, the dolphin needs a streamlined body. It must present the least possible resistence to water flowing along its body, so it can't afford to have big, floppy ears. The external ear of a dolphin consists of a small opening, almost a pinhole. This small opening is covered with fibrous tissue in many dolphins, an arrangement that does not appear particularly conducive to the acute sense of hearing required by these animals.

So the external ear of a dolphin is essentially absent, and the ear canal may not even be functional. The dolphin does have the normal complement of three middle ear bones, however, and like all land mammals, the inner ear contains the cochlea. When we considered the bat's auditory system, we found that the cochlea was encased in the petrous bone, a part of the skull that is especially hard. In contrast, the middle and inner ears of the dolphin are not located in the skull at all. They are encased in bone all right, but the bone is not a part of the skull! In dolphins, the internal auditory apparatus is encased in a separate bone called the bulla. The bulla is joined to the skull via cartilage and connective tissue. Interestingly, it is surrounded by the type of "acoustic fat" found in the melon.

Recall from our discussion of the bat's middle ear that the malleus (or hammer) is the first bone in the ossicular chain that connects the tympanic membrane (eardrum) to the cochlea. In dolphins, the tympanic membrane is not even connected to the middle ear bones! Experiments have shown that removal of the malleus in dolphins results in only a very mild decrease in acoustic sensitivity. In land mammals, the same procedure would result in a thousandfold decrease in hearing sensitivity.

Sometimes people have problems in transmitting sound from the eardrum to the cochlea, and when that occurs, there is indeed a profound loss in hearing sensitivity. In fact, you may at one time or another have suffered from such a hearing loss due to a severe cold. While the hearing loss from a cold generally is only temporary, sometimes the condition can be permanent. If the problem is with the bones of the middle ear, it is called a *conductive hearing loss.* Severe or repeated infections in the middle ear can lead to a stiffening of the ligaments that hold the middle ear ossicles together. As the linkage between the eardrum and the cochlea becomes increasingly inflexible, the hearing loss can be quite severe. In such severe cases, the middle ear bones can be replaced with a prosthetic device. This of course requires removal of the bones. Tests of hearing sensitivity in patients undergoing this surgical procedure have confirmed that hearing thresholds are about 1000 times higher after removal of the middle ear bones. Most of this loss is restored by the prosthetic device.

Now the ears of a dolphin have a dramatically different anatomy than those of land mammals, and the difference raises an interesting question: How is it that acoustic signals traveling in water come to stimulate the dolphin's inner ear? A complete answer to that question is not yet available, but what we do know is quite intriguing. First, as we indicated a moment ago, the malleus does not appear to be very essential, since removing it has very little consequence. The remaining middle ear bones do appear essential for normal hearing sensitivity in dolphins, however, so it is thought that, as in terrestrial mammals, sound vibrations do indeed enter the cochlea via movements of the incus and stapes. Our main problem, then, is to identify how sound energy gets transmitted to the stapes, and then to the cochlea.

We have seen that the ear canal, which is the pathway for conducting sound vibrations to the middle ear of land animals, is probably not functional in dolphins. Why not? Why not use the ear canal? Dolphins have one, but it apparently is vestigial. One reason might be the extreme variations in water pressure that dolphins routinely encounter as they swim between the surface and the ocean depths. Snorkelers and scuba divers are well aware of this problem. And anyone who has flown in an airplane has experienced a milder version of the same effect. The problem is this. If the variations in pressure that we call sound are to move the tympanic membrane back and forth, then the average (or ambient) pressure on both sides of the membrane must be the same. When we are on land, the pressure on both sides of the membrane typically is equal. This equalization of pressure is ensured by a little tube called the *eustachian tube,* which connects the middle ear with the throat. The eustachian tube provides a passageway through which air can pass into or out of the middle ear. This enables changes in the air pressure in the middle ear that match changes in air pressure outside the ear. This is why swallowing, or chewing gum, or yawning can make your ears "pop" when you ride a ski lift to the mountain summit. As we ascend in altitude, the ambient air pressure is reduced. This reduction causes the tympanic membrane to bow outward, since the air pressure inside the middle ear is now greater than the outside air pressure. Swallowing allows the excess pressure in the middle ear to be released into the throat, equalizing the pressure on both sides of the tympanic membrane. When that happens, we hear a "pop," or feel a clearing of our ears.

When we dive underwater, we encounter the inverse problem. Now the ambient (water) pressure is greater on the outside than inside the middle ear, and our eardrum bows inward. This bowing increases the tension on the eardrum, and can produce obvious changes in our hearing sensitivity. Not only is our hearing sensitivity reduced, but as the pressure difference increases, the pressure on the eardrum can become quite painful. The solution? Gently increase the pressure in your throat and nasal cavity until the increased pressure is transmitted throught the eustachian tube into the middle ear. The pain suddenly disappears, and you can hear again!

Imagine, however, the plight of the dolphin. It dives to remarkable depths, comes up for air, and dives again. Over and over. For its whole life. If it had our ears, it would constantly have to adapt to these repeated changes in ambient water pressure. On the other hand, if we could eliminate the air cavity within the middle ear, then the pressure within the ear would automatically change with changes in the ambient water pressure (just as all the other organs of your body do), and the problem would be solved! So it seems likely that the unusual anatomy of the dolphin's ears is an adaptation to the requirements of underwater diving.

In addition, placing the cochlea in the bulla rather than the skull has the effect of acoustically isolating each cochlea from the skull. Since the dolphin's body has about the same acoustic impedance as water, it is essentially "acoustically transparent." That is, sound energy will readily enter the dolphin's body, and very little energy is reflected. This has an important consequence for processing sounds (or echoes) that are off to one side. Typically, the location of these eccentric sounds is detected by comparing the intensity of the signal arriving at each ear. This interaural difference in intensity is especially important in localizing high-frequency sounds, like sonar signals. The ear farther from the source will receive a weaker signal because reflections of the sound energy from the animal's body create a sound shadow. However, if the body has the same acoustic impedance as the transmitting medium (seawater for the dolphin), there is no sound shadow. Placing each cochlea in a separate bone, each of which is surrounded by fat tissue that has a slightly different impedance than the seawater, restores the sound shadow that would otherwise be absent, and aids in sound location.

Well, while this might be an interesting example of how anatomical features relate to functional requirements, it still leaves us with our original questions: How do sounds stimulate the cochlea?

As remarkable as it sounds, the answer appears to be "through the jaw." Dolphins "listen" with their lower jaw! This idea was first suggested by K. S. Norris in 1964. Norris's idea was that sound energy traveling through the water would set up vibrations in region of the lower jaw where the bone is unusually thin. From there, the vibrations could be transmitted to the bulla via the acoustic fat that lies between this specialized part of the jaw and the bulla. These vibrations are thought to vibrate the bulla (remember that it is not rigidly connected to the dolphin's head), and the inertia of the middle ear bones could cause motion of the stapes relative to the cochlea. Thus, relative motion between the middle ear bones and the cochlea would introduce the vibrations into the cochlea.

Hearing through the jaw. Seem a little far-fetched? Well, skepticism is good, but science requires an artful combination of skepticism and open-mindedness. After all, we've seen before that nature can be stranger (and far more clever) than our imaginations; we've learned enough by now to know we can't arrive at the correct answer through intuition. We shouldn't pass judgment without evidence one way or the other. How might we get evidence that dolphins hear with their jaw?

One approach has been to use that time-honored strategy of Lazzaro Spallanazi. We might test animals that have been deprived of the structure that is assumed to be essential. No, no one has removed dolphins' jaws and then tested them on some echolocation task. But researchers have covered the jaw with sound-absorbing material and then tested them for changes in hearing or sonar capacity.

A second approach was devised by Theodore Bullock and his associates at the Scripps Oceanographic Institute in La Jolla, California. Bullock et al. (1968) recorded the electrical activity from a major auditory structure in a dolphin's brain while they presented acoustic stimuli to their subject under a variety of different experimental conditions. In one series of observations, they measured the amplitude of the neural response to auditory stimuli that were presented through a hydrophone pressed against the animal's skin. They recorded how the amplitude of the recorded acoustic response changed as the hydrophone was moved across the surface of the head. What they found was very interesting. The maximal response was obtained when the small underwater speaker was placed near the thinned part of the lower jaw; the region that Norris had suggested could be the receptive site for sound energy! This region of greatest sensitivity to sound waves is illustrated in figure 7.1. Scientists are naturally skeptical people, however, and at least some regarded these results as inconclusive.

More recent work by Brill and colleagues (1988) used behavioral testing of sonar abilities in dolphins fitted with a mask that covered their lower jaw. The animals were also blindfolded, and were trained to discriminate two underwater objects by using their sonar system. Two different masks were used. One was made of a synthetic rubber that was

Figure 7.1
The shaded region illustrates the area where dolphins are most sensitive to sound waves applied directly to the skin. This provided the initial evidence that dolphin sonar reception begins in the jaw, and is then transmitted to the bulla, a bone that is separate from the skull and contains the spiral cochlea.

acoustically "transparent" (it attenuated sound very little), and the other was made of a different synthetic rubber that was acoustically opaque (attenuated sound quite effectively). Sonar discrimination performance while wearing the transparent mask was comparable with results in which no mask was used. In contrast, performance while wearing the opaque mask was significantly disrupted. Although performed 200 years after Spallanzani's initial experiments with masked bats, the similarity of approach is striking. Fortunately, the scientific community was considerably more receptive than they were 200 years ago, and most workers feel the evidence in support of Norris's jaw reception theory is virtually overwhelming.

With some ambivalence, we must acknowledge that little else is known concerning the detailed operation of sonar signal production and sonar signal processing in dolphins. We have alluded to the substantial disparity in the level of understanding that science has achieved about the operation of the bat and dolphin sonar systems. Much of our understanding of bat sonar comes from neurobiological experiments, which require surgical intervention. Neuroanatomical work that establishes the wiring diagram of any neural system requires removal of the brain for microscopic analysis. All of this work is done under the most rigorous standards of ethical and humane treatment of the bats. Similar experiments are simply not done in dolphins. Many would say that they shouldn't be done: that dolphins are too intelligent, too much our equals, and we do not have the ethical right to treat them like laboratory mice or even laboratory bats. There does seem to be a lot of merit to that argument.

In any case, as a practical matter, I can find no reference to experiments that used surgical intervention on dolphins published after the late 1960s, so most modern work on dolphin sonar uses cleverly designed behavioral experimental methods, aided by developments in applied mathematics, digital signal processing, engineering science, acoustics, and noninvasive imaging technologies such as CAT scans and nuclear magnetic resonance imaging (NMRI).

There is essentially no neurobiological experimentation going on with dolphins. So progress is slower and of a slightly different type than what we saw with bats. We do know that dolphins have about the same number of cochlear hair cells as humans, but they have about three times more auditory nerve fibers. We don't know how these fibers respond to auditory signals in general, or to the echoes of sonar signals in particular. It is almost always an interesting exercise to speculate, however. Certainly we can engage in such speculation with little fear of being proved wrong—at least for the foreseeable future. We suspect that the neurons involved in processing dolphin sonar signals are tuned to a narrow range of frequencies that at least follows the pattern of all auditory systems that have been studied using electrophysiological techniques. We might not expect to find the filter neurons so conspicuous in CF-FM bats, however, because the dolphin sonar signal contains no CF component. Dolphins must localize objects and potential prey, so like other animals, they must encode azimuth and elevation of sounds as well as their sonar echoes. Azimuth must still be computed by comparing signals arriving at the two ears, so binaural processing is likely to be an important feature of the dolphin's auditory system. Perhaps the most concise statement we can make is that the structure and organization of the dolphin brain follow "the general mammalian plan," so with respect to the actual processing of sonar signals, we might expect that there are many homologous processes that apply with equal force to bats, dolphins, and even humans.

Despite these expected homologies, we should not lose sight of the fact that bats and dolphins could hardly be more different. The one (perhaps the only) thing they clearly share is their biosonar abilities. Since it is unlikely that these two very different classes of mammals share any but the most ancient ancestors, we strongly suspect that these animals evolved their abilities independently. Biosonar is a wonderful example of convergent evolution: unrelated species evolving a similar ability through independent selection pressures. As such, there is no compelling reason that these systems as elaborated in bats and marine mammals should be the same. Just as the differences we've noted relate to differences in their respective environments, so the similarities probably relate to the most remarkable thing about the evolution of species: that they so often find optimal solutions to problems of adapting to the constraints imposed by the physical world. So the similarities are likely the result of a form of coercion by the physical environment itself.

II

Biological Compasses

8

Maps, Mobility, and the Need for a Compass

Have you ever wandered through a large, labyrinthine building and suddenly realized that you are completely disoriented? That you don't know which exit will take you to your car or the proper subway station? Even though you know the downtown area well, and you know you came in the south entrance, after a number of twists and turns and several elevator rides, you have no idea, as you walk down yet another corridor, whether you are walking north, south, east, or west?

In some buildings, the ability to produce spatial disorientation in visitors appears to have been one of the architect's criteria for good design. It seems to me that many modern hotels are like that. One that stands out in my mind is the Westin Bonaventure Hotel in Los Angeles. It's one of those big, glittery, multitowered buildings. There is the Central Tower, which is surrounded by four smaller towers: the Yellow, Green, Red, and Blue towers. Quite impressive when seen from the outside. And from the inside, too. The Los Angeles Bonaventure has one of those incredibly spacious lobbies where the ceiling seems five or six stories high. The cantilevered balconies, large plants, shopping arcades, sidewalk cafés, piano bars, and elevators in those glass tubes all help create the illusion of being outdoors.

As a guest at the Westin Bonaventure, I had only one problem with this very beautiful hotel. It was so easy to get lost! There are four sets of elevators, and you have to use the right elevator to get to your room (figure 8.1). Sure, the elevators are color-coded. But the signs aren't that

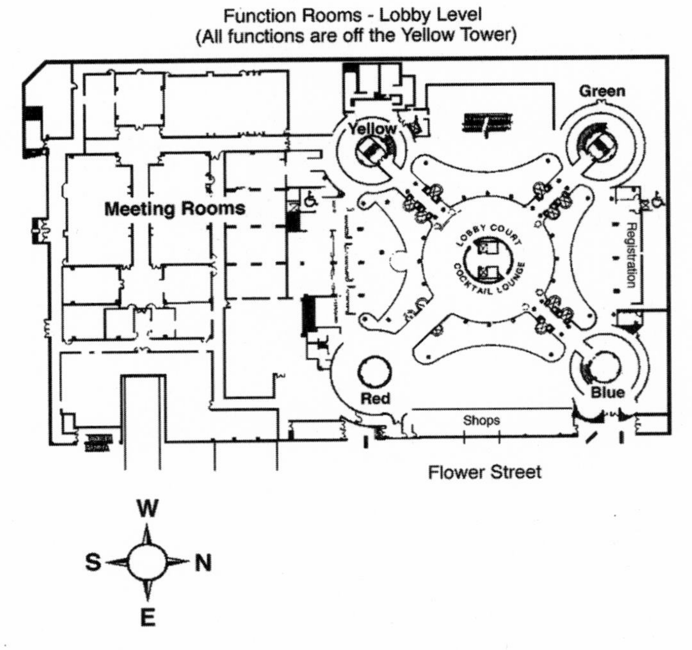

Figure 8.1.
Schematic illustration of the lobby of the Westin Bonaventure Hotel, 404 South Figueroa Street, Los Angeles. The lobby is quite large, and it is easy to get lost in it. It is probably more precise to say that one is readily disoriented, since you know all along that you are indeed in the hotel lobby! From the program of the 36th Annual meeting of the Psychonomic Society, 1995. Ironically, the program organizers provided a compass rose with the map of the lobby. This would be of help only if you brought your compass with you, or if you could otherwise sense compass directions.

big, and if you are nearsighted like me and frequently forget your glasses, it is hard to look around the incredibly spacious lobby and find the little signs for your elevator. So it just seems easier to keep walking from one elevator to the next until you find the right one. When you've had your last drink at the piano bar, or just come out of a conference session, or finished your conversation with a fellow conferee, you have a compelling illusion that you *know* where the "blue" elevator is. So you begin your trek across the lobby, but alas, when you get there, it turns out you've arrived at the red elevator, not the blue. Now which way? You think it's over here. You walk past the café, past the piano bar (again) and . . . no, that's not it either. That's the green elevator. Funny, it looked blue from way over by the red one. . . . Wait, it must be over that way. Finally! You've found the elusive blue elevator.

It seems that most of us are not that good at navigating in unfamiliar environments. Of course, in a day or two you get the hang of the place. But by then, your convention is over and it's time to go home! Next year, the meeting will be in a new maze. Maybe it is true that some people have a better sense of direction than others, but even the best of us get confused. My brother Mark is a case in point. He likes to refer to himself as the "human compass." And he is pretty good (although in a pinch, I'd rather have a real compass). Mark's the kind of person who can readily find his way home after visiting in an unfamiliar neighborhood. When you come to an intersection, and realize that you've lost track of the return path and don't know whether you should turn left or right, Mark usually knows which way is the way home. Mark likes to think that if Hansel and Gretel had been more like him, they wouldn't have needed to rely on a little trail of crumbs. But even the human compass gets disoriented.

A case in point. Once Mark and a friend were scuba diving in the Florida Keys. They had rented a boat, and had gone about 5 miles offshore to explore a coral reef that was said to be particularly beautiful. Enthralled with this undersea world, the divers meandered through the coral, admiring the colorful fish that populate the reef. When they were getting low on air, they decided to surface. It was then that they realized there was a problem. They couldn't find the boat! They had committed one of the cardinal sins of offshore scuba diving: they had wandered so far that the boat was now out of sight. What to do? Which way should they swim? Allan felt strongly that they should head in one direction. Mark felt equally sure that they should head in the opposite direction. Well, as Mark tells the story, they followed his instincts, and after swimming for a tense 20 minutes or so, found the boat. The human compass had prevailed again. Was it skill or luck? No matter, the point is made—for all their talents, humans are not particularly well endowed with an absolute sense of direction or navigation ability. If we were, there would have been no need to invent compasses, sextants, or GPDs (global positioning devices). GPDs use a system of satellites in geosynchronous orbits to pinpoint your position anyplace on earth with an accuracy of 10 meters!

On the other hand, we are all familiar with tales of guides who know a particular region "like the back of their hand." While certain of the more astonishing feats of human navigation may be apocryphal, it is certainly true that people can, through experience, generate incredibly detailed "internal maps" of their surroundings. In many such cases, the landscapes look entirely uniform to an outsider: an endless expanse of desert, a dense tropical rain forest, or a grass-covered plain. But the landmarks are still there, at least to the trained eye.

These internal maps are often called "cognitive maps," and they almost surely rely on knowledge of important landmarks, the approximate distances between them, and their relative directions. These directions need not relate in any way to compass directions (north, south, east, or west), but certainly they can. It is possible to generate cognitive maps without regard to compass headings. Many people may know their way around a city, know how to get from point A to point B, but not nessarily know that point A is northwest of point B, or that Front Street runs north and south, while Market Street runs east and west. All that is really needed is knowledge of one's current position and destination within the cognitive map. Then it is simple to plan a path going from A to B, using the internal map, and verify progress by noting the passing of expected landmarks along the way.

In other situations, possessing a cognitive map may not be needed to get where you need to go. For example, a migratory animal living in the northern hemisphere may be able to find its winter home simply by heading south. While a cognitive map may not be required in such a situation, knowledge of compass directions obviously is.

These considerations lead us to two useful concepts related to the guidance of migratory movements and other forms of animal navigation. *Orientation* refers to the ability to identify different compass directions, and *navigation* usually refers to an ability to identify one's current position and move to some new location. Orientation, then refers, to an ability to gain an internal representation of compass directions. Navigation implies knowing where you are and where you want to go. It may or may not make use of orientation abilities. My brother and his friend could have found their boat using either method, or a combination of the two.

Many animals are faced with the need to traverse incredibly long distances. Often this need is produced by seasonal changes in climate or by scarcity of food supplies. Animals that prey upon migratory animals may themselves have to migrate. Consider the bobolink, a small bird that spends summers in Canada and the United States, and winters in Brazil, Bolivia, and Argentina. Each fall, the bobolink must fly from its summer range in North America to its winter range in South America, a distance of up to 7000 kilometers!

Many migratory animals make journeys at least that long. In fact, the world record holder is the arctic tern, which migrates between the North and South poles, the longest possible distance on the planet. It seems particularly ironic that this longest of all animal journeys always ends at a polar ice cap! Most of us would want to see some palm trees and feel a gentle tropical breeze at the end of such a long journey. And we wouldn't need an elaborate sense of direction to know, when the destination looks so much like the place we had just left, that we had gone too far.

Birds are not the only animals that migrate with the seasons. Many marine mammals migrate between arctic and tropical waters, and most of us are familiar with the remarkable ability of salmon to return to the exact stream where they were hatched, after several *years* of swimming in the open ocean. Even some insects migrate. These itinerant arthropods raise a question that has puzzled humans for centuries: How do migratory species know how to get between their winter and summer homes? If people can get lost in a large department store, why don't monarch butterflies get lost? Their journey from the eastern United States to Mexico takes several generations to complete! Those that begin the expedition never reach the destination. Do these animals rely on orientation alone, or do they navigate using cognitive maps? Is it even sensible to wonder whether an animal possessing a nervous system as small and primitive as an insect's could create or be genetically endowed with maplike representations of the world—even when our own

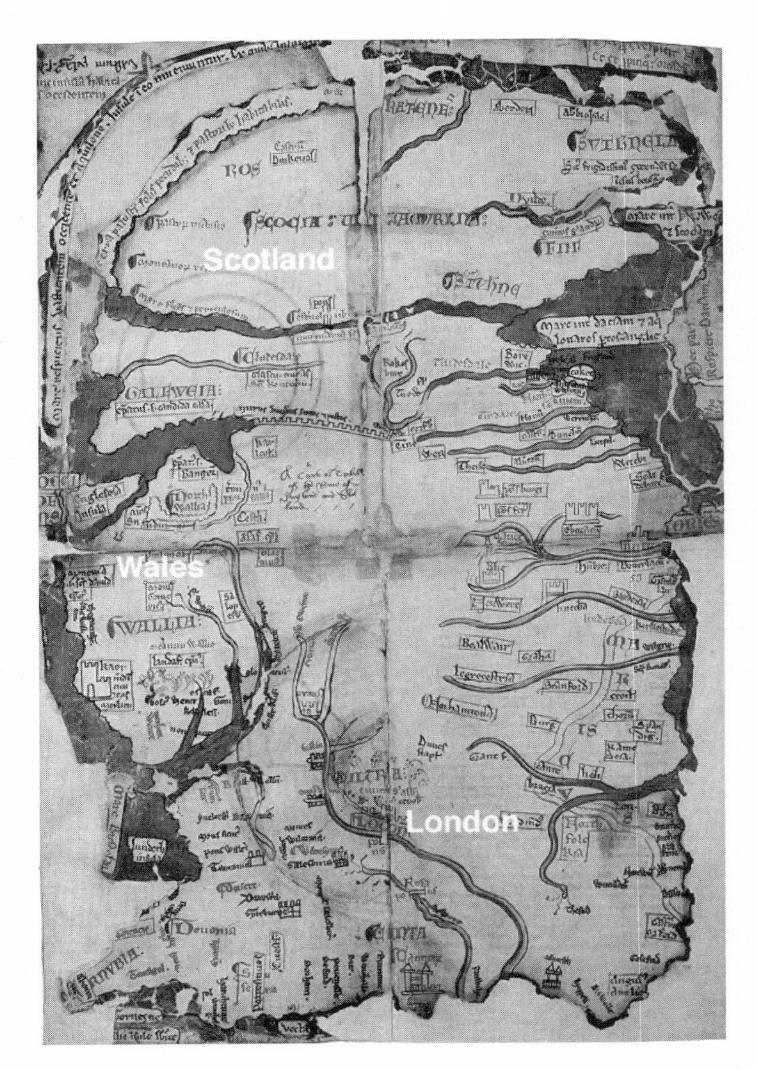

Figure 8.2
A map of Great Britain circa 1250. I have added some landmark designations to help orient the reader who is well acquainted with the actual geography of Great Britain. It was another 300 years before the science of cartography had developed the methods needed to produce a geographically accurate rendition.

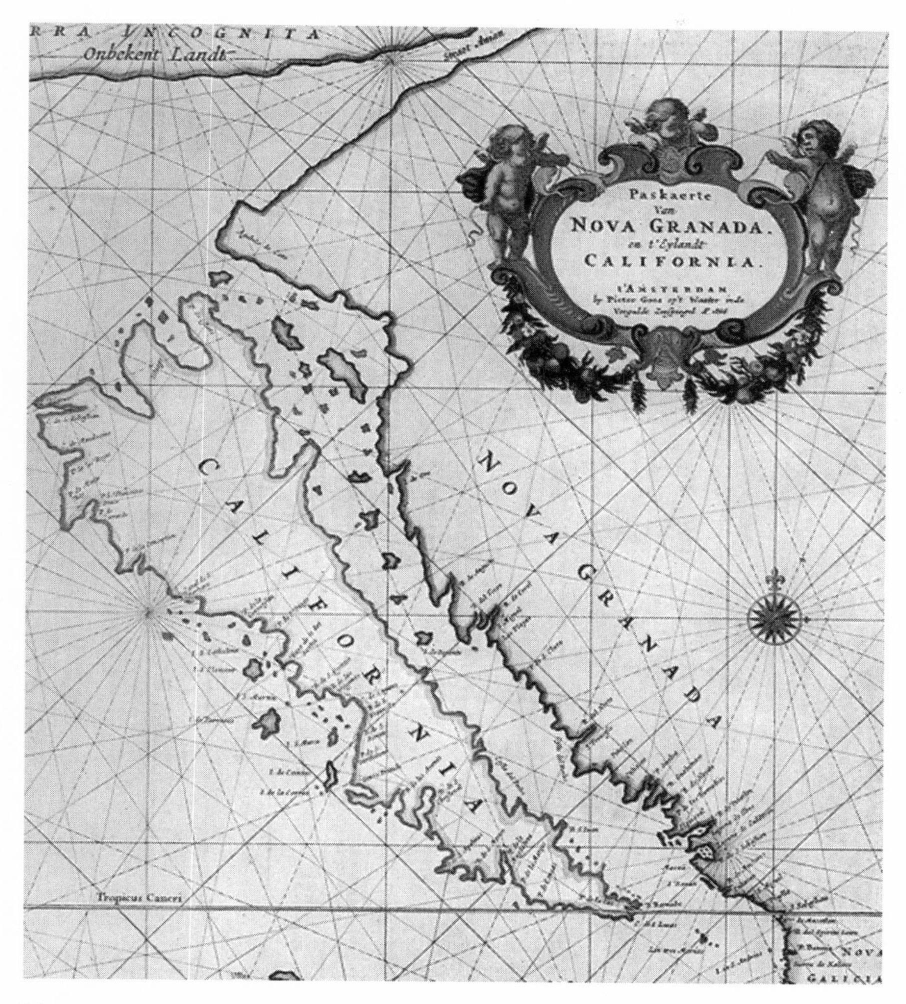

Figure 8.3
Map of the "island" of California, 1666. Can you imagine trying to find California by land, using this map? After a cross-continental journey, you would reach a body of water and sail westward. Expecting to find land, you would instead find 3000 miles of Pacific Ocean.

representations were so woefully inadequate as little as several hundred years ago? (See figures 8.2 and 8.3.) We shall see that while answers to these questions are beginnining to be revealed, many mysteries remain. Above all, we shall see how, once again, the systematic creativity of science is beginning to unveil the astonishing creativity of life on Earth.

One important aspect of the creativity of animal orientation and navigation is the use of multiple cues. As in a modern airplane, backup systems are essential. Reliance on only one system is risky, even dangerous. Earth's surface contains deposits of magnetic rocks that can produce substantial local distortions in Earth's magnetic field, so exclusive reliance on a magnetic compass could produce problems. As every outdoors person knows, the position of the sun in the sky can indicate compass direction, but this requires knowledge of the time of day. And in any case, the sun isn't always visible. For centuries, sailors have used the stars as an aid to navigation, but that information is not available during the day or during dense cloud cover. Celestial and solar cues are unlikely to be particularly useful to fish and marine mammals in any case. Thus, we shouldn't be surprised to find that migratory abilities are based on multiple sources of information.

It is also true that orientation alone is probably not sufficient to account for the success of long migrations. During migratory flight, crosswinds can blow the birds off course, even if their compass heading remains constant. The same applies to marine animals: ocean currents make it impossible to find a given location based on compass orientation alone. So landmarks or other maplike representations are likely to be an important source of information for animal navigation.

Any complicated system that is controlled by a wide variety of input cues is difficult to study scientifically. The problem is that, in experiments, we like to keep all variables constant except for one, which the experimenter systematically manipulates. This way, any change in the output or performance of the system can logically be attributed to the experimental manipulation. This technique has been used to great advantage in experimental work on animal navigation. However, it does require that other potential cues be eliminated (or at least held constant). Using this single-variable method, one can certainly determine whether an animal is sensitive to certain navigational cues. It is a much more difficult matter to determine how these cues normally work together in a coordinated fashion to permit migrations across oceans, continents, and hemispheres. So while we don't yet have a completely integrated picture, we do have glimpses of the kinds of processes that permit animals to perform their astonishing feats of navigation. And those glimpses, as incomplete as they might be, are really quite remarkable.

9

Animal Migration: A Compass in the Head?

For centuries humans have known of and taken advantage of the remarkable ability of homing pigeons (they apparenty were first used to carry messages by the Romans 3,000 years ago), and as early as the 1850s a German named Alexander Theodor von Middendorf suggested that many birds might possess a magnetic compass. But the scientific study of animal navigation is usually said to have begun in the years between 1958 and 1965, when a group of zoologists led by Friedrich Merkel, a professor at the University of Frankfurt in Germany, made a startling and initially quite controversial discovery. Merkel and his colleagues, H. G. Fromme and Wolfgang Wiltschko, reported that the "premigratory restlessness" of the European robin was generally oriented in the appropriate migratory direction, even when the birds were placed in an enclosed room that precluded any visual cues. The implication was that these birds must be able to sense compass directions. The orientation of the bird's activity is very noisy—they hop around in all directions. The orientation bias is evident only following a statistical analysis. This variability in the direction of premigratory restlessness was one of the reasons that Merkel's initial findings were met with a great deal of skepticism. But Merkel's experiments included one important innovation—an objective method for measuring the bird's activity over long periods of time. The method permitted an objective analysis of the results, and could readily be implemented by other scientists wishing to replicate the experiment. Following a series of successful replications, most experts on bird migration came to the conclusion that migratory birds do indeed possess a sensitivity to Earth's magnetic field.

Merkel's innovative apparatus was, by today's standards, a relatively simple affair. It consisted of an enclosed octagonal chamber. A perch was positioned on each wall of the chamber, and when the bird hopped onto the perch, its weight activated an electrical switch. By counting the switch closures on each perch, the distribution of hopping orientation could be computed. When Merkel and his colleagues analyzed the frequency with which the robins hopped onto each perch, they discovered that the birds hopped onto each perch, and there was little evidence that one perch was favored over any other. However, when the average direction for each night was computed, and the means over many nights were taken together, a pattern of nonuniformity emerged. This is how the bias in hopping was discovered.

Interestingly, the bias is directed toward the south in the fall and toward the north in the spring—the seasonal migratory directions. What was even more startling was that when the orientation cage was shielded from the Earth's magnetic field, the hopping orientations became random!

So it appeared that the birds could sense the geomagnetic field. Skeptics argued that the bias was weak, and that small variations in the sensitivity of the electrical switches could have produced spurious results. This objection was countered by a clever modification in the design of the orientation cage by S. T. Emlen, a member of the Frankfurt group. Emlen used a circular chamber with a funnel-shaped floor. An ink pad is placed at the base of the funnel. Since this is the only place that the bird can comfortably stand, the arrangement ensures that the bottom surfaces of the bird's feet are covered with the ink. The birds are housed in this cage each night, and when they are active, they jump onto the funnel-shaped part of the floor. The funnel portion is lined with paper, so that each jump leaves an ink mark on the paper. The ink marks provide a record of the activity pattern over the course of the night. Figure 9.1 shows a diagram of the Emlen funnel.

Although it is an exceedingly simple device, the Emlen funnel revolutionized research on the cues that enable orientation in migratory birds. For the first time, the importance of celestial cues, the sun, and Earth's

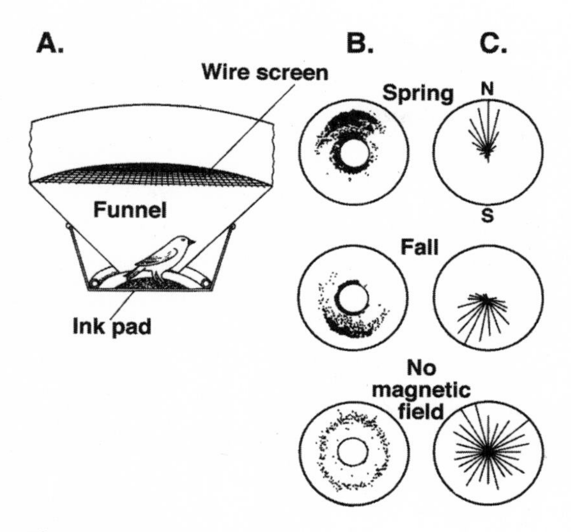

Figure 9.1
A. The Emlen funnel, a device that is used to measure the orientation of hopping in birds experiencing premigratory restlessness (also called *Zugunrule*). The pattern produced by the hopping is shown in B. C illustrates how the number of footprints in each direction can be represented as vectors, the length of each vector representing the number of hops in that direction. These vector diagrams permit easy visualization of the results. The direction is biased toward magnetic north (N) in the spring and magnetic south (S) in the fall. When the cage is placed in a magnetically shielded chamber, the orientation bias disappears.

magnetic field could be evaluated under controlled conditions. The logic behind these experiments was the same as that used by Spallanzani. If the orientation biases could be observed when a particular cue was not available, then that cue was not essential for orientation. If, on the other hand, orientation biases were observed when the birds were selectively exposed to a particular cue, the birds must have a means to detect and use that particular source of information.

To be sure, there were still problems. The first was a purely practical issue. Since the premigratory activity occurred only in the fall and spring, data could be collected only twice a year. This places serious demands on the patience of the experimenters. If a given set of results turns out to be ambiguous, or suggests an interesting new experiment, one has to wait at least six months to resolve the ambiguity or test the new idea.

The second problem is a little more difficult, and relates to the type of statistical analysis we've seen is needed in order to reveal the orientation bias. The procedure of getting a mean direction for each night's hopping, and then computing the mean for many different nights is called a second-order analysis (a mean of means, if you like). In a second-order analysis, each bird produces one data point per night. You need to test the same bird over many nights to get one bird's orientation bias, and a well-designed experiment is going to require the use of many birds. So data collection is a time-consuming task.

An example of typical results in an Emlen funnel is provided in figure 9.1. The dispersion of the individual birds' results is substantial. It would be difficult to lower your handicap in golf if your tee shots showed this kind of scatter! These biases, however subtle, are consistently found in many species, and they are the basis for much of what is known about orientation mechanisms in migratory birds.

Work using the Emlen funnel has shown that the birds do not rely on a single cue. Their orientation behavior can be guided by the sun, the stars, and Earth's magnetic field. Let's consider the kinds of results that lead to this conclusion.

The Discovery of a Solar Compass

We are all familiar with the fact that the sun rises in the eastern sky and sets in the western sky. It's been this way for quite a long time, and many species have caught on to the fact that the sun provides one cue to compass direction. It is also true, however, that accurate use of a "sun compass" requires knowlege of the time of day. To take an extreme example, if we thought that it was dawn when is was really dusk, then we would mistakenly judge the western sky as being in the east. In this case, all orientation behaviors would be in error by 180°. There is nothing exotic about the sensory modality that permits use of a sun compass, however; it is clearly vision.

Figure 9.2 depicts the results of the type of experiment that demonstrates the use of a sun compass with European starlings. Starlings were placed in an Emlen funnel located in a circular room. Sunlight entered the room through windows and, as illustrated in the diagram, it came

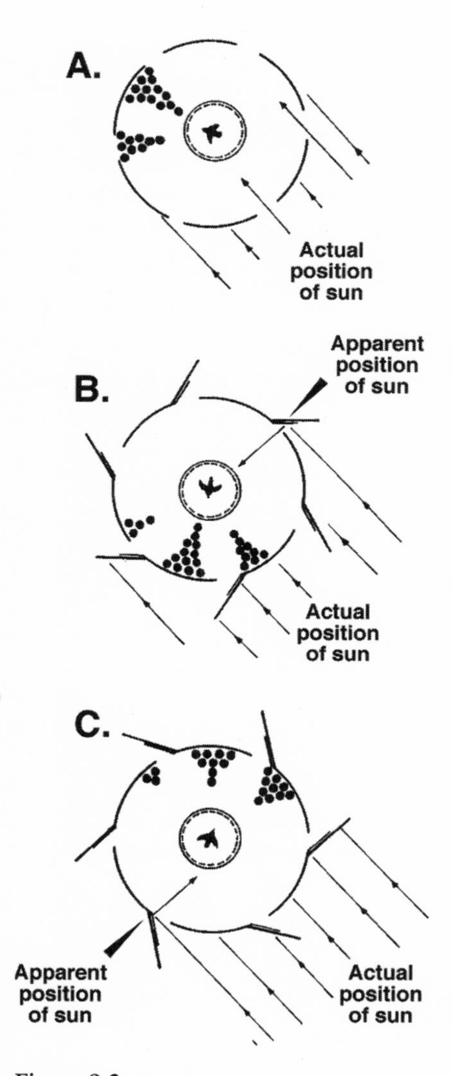

Figure 9.2
Demonstration of a sun compass with European starlings. The experiment was performed by Gustav Kramer in 1950. In A, the direction of migratory restlessness of starlings was shown to be oriented opposite the location of the sun. In B and C, mirrors were used to change the apparent location of the sun, and the orientation of the birds' activity changed according to the apparent direction of the sun.

from a single direction. Under these conditions, premigratory activity was oriented toward the northwest. In a second series of observations, the direction of sunlight entering the chamber was altered by an arrangement of mirrors, which caused a change in the apparent position of the sun. When the artificial "sun" was shifted with respect to the apparatus by 90° (i.e., so that the sun appeared to be in either the northeast or the southwest, the bird's orientation activity shifted by a corresponding angle. Thus, the orientation of the activity appeared to be determined by the position of the sun.

To establish that the starlings possessed the internal clock required by use of a sun compass, the birds were trained to feed from one container in a circular cage that had 12 identical containers along its circumference. They were trained to feed from the northernmost feeder under a normal sunlit sky. Despite the fact that the position of the sun changed relative to the critical feeder, the birds learned to sample from that particular feeder. In effect, they learned to go to the "northern feeder," presumably using their sun compass to determine which feeder faced north. When the same birds were exposed to an artificial light-dark cycle that was 6 hours behind (earlier than) the natural light-dark cycle, they sampled from a feeder that was shifted 90° from north (the easterly direction). Thus, the birds shifted their interpretation of compass direction based on a change in the phase of their internal clock. This sort of observation indicates that the birds can determine compass bearing by using the position of the sun in conjunction with the time of day, as accurate use of a sun compass requires. It also suggests that when information about the position of the sun is available, that information dominates any the bird might have regarding magnetic compass direction.

A Celestial Compass

Similar experiments have shown that some birds possess a celestial compass, an ability to orient along compass directions using the stars. Using a celestial compass is somewhat more complicated than using the sun compass. When not obscured by cloud cover, the sun is by far the brightest object in the sky. So it doesn't take a sophisticated visual system to detect it. Stars are much dimmer objects, and they are distributed in patterns (i.e., the constellations). So a compass system based on the stars requires a more elaborate visual system. This does not seem to be a large obstacle, since the visual system of most birds is actually quite good. Most birds have excellent acuity, and they see in color. So the avian visual system is up to the task of analyzing the position of the stars.

Like the sun, stars move through the sky over time, so a celestial compass that relies on one star, or perhaps one pattern of stars (like a constellation), would also require an internal clock. The clock is needed because the position of a star or constellation does not unambigously indicate compass direction unless the time of night is taken into account. Work on the sun compass clearly indicates that the animals have such a clock, but evidence in ducks indicates that while they have an internal solar clock, they do not have an internal stellar clock. That is, while the ducks do show a sensitivity to the day–night cycle when orienting in the day, they do not make similar adjustments at night.

While this may at first seem puzzling, one important aspect of the motion of the stars through the nighttime sky may be the key to the dilemma: *all the stars revolve around a relatively fixed point.* As every Boy Scout knows (or at least once learned), the center of stellar rotation in the northern hemisphere corresponds to Polaris. Because Polaris remains in a fixed position in the northern sky, it is called the North Star. If birds could identify Polaris, they would have a fixed reference point to indicate the northerly direction. Many residents of the northern hemisphere have learned that two of the stars in the constellation known as the Big Dipper always point toward Polaris, so if you can find the Big Dipper, you can find Polaris, and thereby establish compass headings (figure 9.3). Is it possible that birds have somehow internalized the same information?

Recent experiments suggest that they have. It was already known that birds could successfully orient at night when the sky is clear. This had been shown using the Emlen funnel, but it was also shown using homing pigeons that were taken in cages to a novel release point and allowed to fly to their home roosts. Many observations indicated that the pigeons would accurately fly in the direction that would take them home, even though they had been deprived of any visual input during their outward journey to the release point. Imagine whether you could do this. The

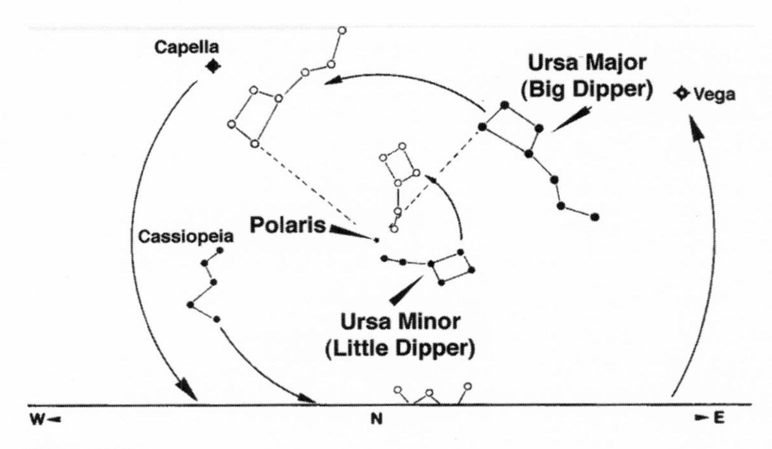

Figure 9.3
In the northern hemisphere, stars appear to rotate around the polestar, Polaris.

analogous situation would be to blindfold a person, take him many miles from home, release him in a field he had never visited before, and see if he could find his way home. A neat trick. Almost no one could immediately begin the return trip by walking in the appropriate compass direction, but homing pigeons can do just that. Many thought they must be able to get a fix on their current position by using the stars, because when they were released on nights with heavy cloud cover, their initial directions of flight were much more scattered than on clear nights. Homing pigeons can also find their way home on cloudy nights, and we will address the mechanisms that support orientation in these conditions shortly. For the moment, we simply wish to show the kind of observations that seemed to strongly suggest that many birds do indeed possess a celestial compass.

In order to see how the celestial compass works, we turn to experiments that were done by Professor S. T. Emlen using indigo buntings. Imagine that we have some newly hatched buntings. Imagine, further, that we rear these young birds under carefully controlled conditions. More specifically, the birds are reared in cages that are placed beneath an artificial sky. The birds are in a planetarium. The planetarium can produce any arbitrary pattern of stars in the nighttime "sky." We expose some birds to the pattern of stars that is naturally seen in the northern hemisphere. That is, all the stars revolve around Polaris. For another set of birds, we provide an artificial pattern of motion: all the celestial bodies revolve around a different star, Betelgeuse. Betelgeuse is the type of star astronomers call a red giant. It's located in the constellation Orion. If you can find the constellation Orion, you can see that Betelgeuse actually has a red tint.

After exposing each set of birds to a different pattern of celestial rotation, Emlen tested each set of birds in—what else?—an Emlen funnel. The orientation of premigratory activity in both groups of birds was tested with a *stationary* sky and measured. The first set of birds—those reared viewing the natural sky—oriented toward the south, the appropriate direction. However, those that had had experience with the stars rotating around Betelgeuse treated it as if it were the North Star! Their premigratory activity was correspondingly misdirected.

When you think about it, this is a pretty remarkable finding. First, it implies that birds are endowed from birth with an ability to obtain a celestial reference point from the pattern of stellar rotation. It would seem that it is in their genetic makeup to to this. The genes, however, do not specify which star should be treated as the North Star, only that the center of rotation should be treated as an indication of north. The actual selection of the reference point is dependent on the early experiences of the birds.

The Discovery of Magnetoreception

The Chinese used magnetic compasses as early as the second century A.D., but they were not used in the Western world until a thousand years later. It was not until 1859 that the idea that animals might be able to detect some aspect of Earth's magnetic field was first suggested in the scientific literature (Middendorf, 1859). The idea was met with great skepticism, which to some extent persists today. However, the persistence of many investigators, perhaps most notably Friedrich Merkel and his students at the University of Frankfurt, has amassed an impressive body of evidence suggesting that migratory birds do indeed sense some aspect of Earth's magnetic field, and use this ability to navigate accurately over long distances.

The evidence for *magnetoreception* has been primarily behavioral in nature, but has come from a variety of sources and a variety of animal species. Not only birds, but also bees, ants, and marine animals such as turtles, salmon, sharks, and whales are suspected of using a magnetic sense as an aid to orientation.

The initial evidence for magnetically based orientation in migratory birds came from a classic experiment by Merkel and Fromme in 1958. The experiment was both elegant and conceptually simple. The investigators simply placed European robins in an Emlen funnel that was devoid of orientation landmarks. The idea was to deprive the birds of a solar compass, a celestial compass, or any other orientation cue. Despite this apparent lack of external cues, Merkel and Fromme observed the now familiar bias in premigratory orientation activity. Beyond that, they observed (Fromme, 1961) that the orientation bias disappeared when the Emlen funnel was placed in a steel drum that substantially reduced the intensity of the geomagnetic field. The implication seemed clear: the birds must be able actually to detect Earth's magnetic field, and the orientation bias observed under these conditions must be based on this magnetic sense.

Merkel and his students were not content to stop there. Merkel and one of his most successful students, Wolfgang Wiltschko, wanted to know what *kind* of compass the birds were using. They noted that there were two features of Earth's magnetic field that might provide the magnetic cue. These two possible cues are illustrated in figure 9.4. The figure shows the fact that Earth is an enormous magnetic dipole: it has a north and a south magnetic pole. This dipole creates a three-dimensional magnetic field in which lines of magnetic force originate from the south magnetic pole, exit from the surface, extend into the atmosphere, and reenter the surface at the north magnetic pole. There are two distinct features of these lines of magnetic force. One is the *polarity;* that is, all the lines of force have a direction, which is indicated by the arrows. By arbitrary convention, the arrows always point toward the north magnetic pole. The second feature corresponds to the orientation of the lines relative to the force of gravity, and is called the *inclination.* Inspection of figure 9.4 illustrates that the lines of magnetic force are parallel to Earth's surface at the magnetic equator (where the inclination is said to

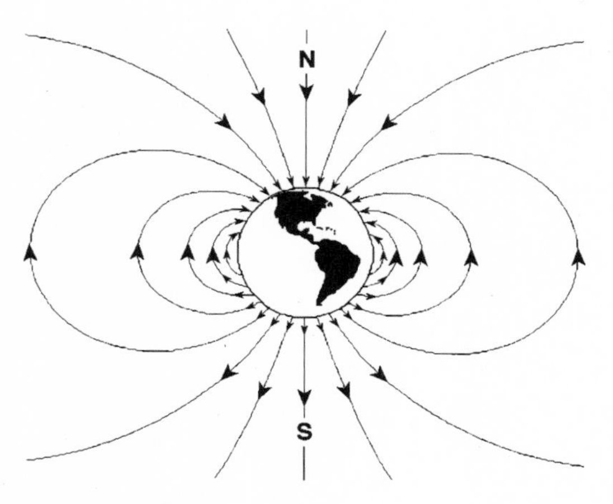

Figure 9.4
Schematic diagram of Earth's magnetic field. By convention, lines with arrows are used to represent the lines of magnetic force. By the same convention, the direction is from the south magnetic pole to the north. The figure illustrates that the lines of magnetic force enter Earth's surface at angles that depend on the distance from the magnetic equator. These angles are called angles of inclination. The angle of inclination at the magnetic equator is 0°. As you move toward either magnetic pole, the angles of inclination (the angle between the lines of magnetic force and Earth's surface) gradually increase until, at the magnetic poles, the angle of inclination is said to be 90°. The key point is that an animal might be able to sense its latitude if it could detect the angles of inclination in the geomagnetic field.

be 0°), and are perpendicular to the surface at the magnetic poles (where the inclination is said to be 90°). The angle of inclination gradually changes with increasing latitude (distance from the magnetic equator).

These lines of magnetic force can be thought of as being composed of two components: one is horizontal and the other is vertical. In mathematics, these components are called *vectors*. The actual inclination at the surface is a combination of the horizontal and the vertical components, as illustrated in figure 9.5. At the equator, the horizontal component is quite large, and the vertical component is 0. The situation at the magnetic poles is reversed: the vertical component is large and the horizontal component is 0. Everywhere else both vectors contribute to the observed inclination. A mathematical technique called vector summation specifies how these component forces are combined.

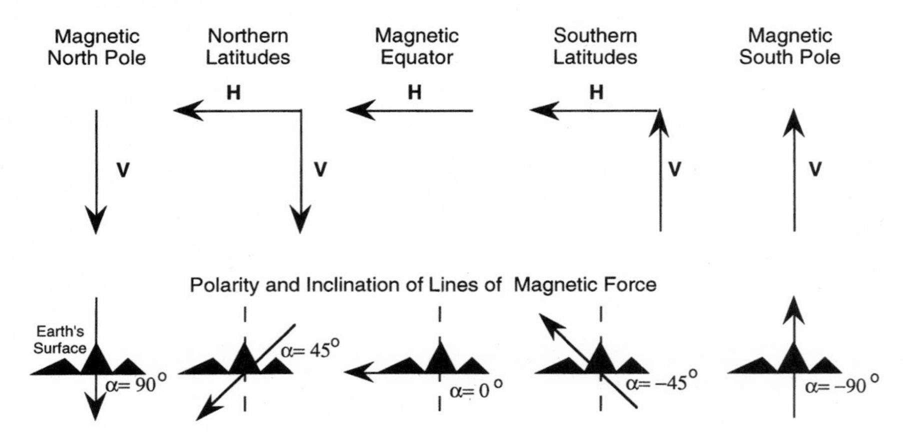

Figure 9.5
The lines of force in the geomagnetic field can be considered combinations of a vertical component (V) and a horizontal component (H). The magnitude of the horizontal component is zero at the magnetic poles, and is largest at the magnetic equator. In contrast, the vertical component is largest at the poles and zero at the equator. These components add to produce different inclinations at different latitudes, as illustrated along the bottom row of the figure.

If a small magnetic needle is mounted in such a way that it is both free to rotate and free to change its inclination, it will align itself with the lines of magnetic force. Not only will the needle point toward the north magnetic pole, but it will align itself with the different angles of inclination at each latitude. At the equator, the needle will lie parallel to the ground but point north. In the northern hemisphere, the needle will still point to the north, but the north-seeking end will also point toward the ground—at increasing angles as you approach the northern magnetic pole. In the southern hemisphere, the north-seeking end of the compass needle will point toward the sky. The only reason that your compass doesn't tilt in this way is that compass needles normally have small weights placed on them to offset the inclination angle. As a result, many compasses work well only within a certain range of latitudes.

Wiltschko recognized that, in principle, the birds might be sensitive to either the polarity or the inclinations of the geomagnetic field: in effect, the birds could possess either a *polarity compass* or an *inclination compass*. We are all familiar with a *polarity compass*, the kind that always

Figure 9.6
Illustration showing the angle of inclination (α) at two different latitudes: one in the northern and one in the southern hemispheres. An animal that possesses an inclination compass can tell which direction is "poleward," but can't discriminate magnetic north from magnetic south.

aligns itself with the magnetic poles. The needle of a polarity compass is mounted such that it is free to rotate in just one plane. It cannot change its angle of inclination, so it can orient only to the polarity of the magnetic lines of force. In effect, a polarity compass is sensitive only to the horizontal component of Earth's magnetic field.

In contrast, an *inclination compass* responds only to the angle of the magnetic field relative to the force of gravity (i.e., the inclination). The "needle" of an inclination compass points toward the nearest magnetic pole at an angle that corresponds to the angle of inclination at that particular latitude. At the equator, the needle is horizontal. As you move toward either magnetic pole, it points downward (toward the ground), in the direction of the pole. Thus, an inclination compass can indicate only which direction is poleward; it gives the same reading at similar latitudes in both the northern and southern hemispheres. However, an inclination compass can also provide information about latitude, since inclination varies (more or less) systematically with latitude. The essential ambiguity of an inclination compass is that it points to the North Pole when you are in the northern hemisphere, and points to the South Pole when in the southern hemisphere. Thus, in order to use an inclination compass as a guide to navigation, you have to know which hemisphere you are in.

Wolfgang Wiltschko and colleagues performed an ingenious series of experiments in an attempt to determine which type of compass was mediating the orientation behavior of their birds (Wiltschko and Wiltschko, 1972). They made use of the intimate physical relationship between electric fields and magnetic fields that was described by the great British physicist James Clerk Maxwell. In short, Wiltschko and Wiltschko produced their own magnetic fields by passing electric currents through large rings of coiled wire called Helmholtz coils. An appropriate arrangement of such coils permitted the investigators to independently reverse the vertical component, the horizontal component, or both components of the magnetic field around an Emlen funnel. Thus, birds could be tested under conditions in which an artificial magnetic field produced a simulated reversal in the polarity, the inclination, or both the polarity and the inclination, of Earth's magnetic field.

The arrangement is illustrated in figure 9.7, which also shows how these manipulations can identify whether the magnetic sense is based on

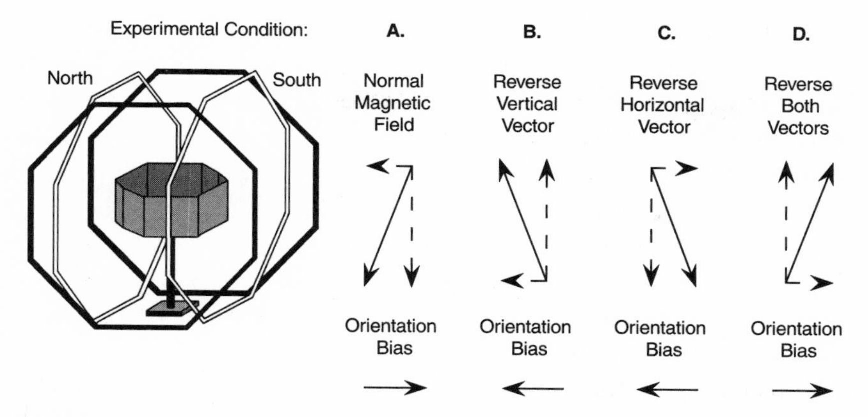

Figure 9.7
An illustration of the experimental arrangement used to determine the nature of the biological compass possessed by migratory birds. The experimental manipulations and the results are depicted in panels A—D. When the lines of magnetic force were oriented as in A and D, the birds (robins) reacted as though the magnetic pole was toward the left in the figure. When the lines of force were oriented as in B and C, their orientation behavior was shifted by 180°. These results indicate that the birds oriented according to the inclination of the lines of magnetic force (which are the same in A and D), rather than the polarity (which are the same in A and B)—that is, they possess an inclination compass.

an inclination or a polarity compass. In panel A, we see a representation of the inclination and polarity of the geomagnetic field corresponding to the latitude around Frankfurt, Germany. In B we see how appropriate voltages applied to two Helmholtz coils will reverse only the vertical component, producing an artificial magnetic field that has the same polarity as in A, but a reversed angle of inclination. In C, we show the arrangement that produces a reversal in only the horizontal component, and in D both components are reversed. The readings from a polarity compass are reversed anytime the direction of the magnetic vector is reversed (e.g., from A to C or A to D). The reading from an inclination compass is reversed when the slope of the line reverses independent of the vector's direction (A to B or A to C). The readings of both an inclination compass and a polarity compass are reversed when only the horizontal component is reversed (A to C).

So we see now how the experiment can determine which type of compass the birds possess. In the normal geomagnetic field, the orientation bias in the fall is southward. If the birds' magnetoreception is based on a polarity compass, this bias should be reversed (oriented northward) in conditions C and D, but not B. If they possess an inclination compass, the orientation bias should be reversed in conditions B and C, but not D. So conditions B and D are the critical conditions. All the work to date indicates that reversals of the premigratory orientation bias occur for condition B, but *not* for condition D. The remarkable conclusion is that the birds possess an inclination compass, not a polarity compass. When exposed to an artificial magnetic field that is purely horizontal (like what is encountered at the equator), the birds show no orientation bias, which is the outcome expected only from an inclination compass. Such findings have been obtained not only in European robins but also in several other species of migratory bird, including both Old World and New World species. The original findings of Wiltschko and Wiltschko have been replicated in several other laboratories.

We indicated earlier that there are seemingly both advantages and disadvantages to an inclination compass. One apparent advantage is that the angle of inclination can provide information concerning latitude as well as compass direction. We also pointed out some disadvantages. First, an inclination compass is useless at the magnetic equator, and second, in order to fix a compass direction (N, S, E, or W), you have to know whether you are in the northern or southern hemiphere. Many species confine their migrations to either the northern or the southern hemisphere. Since they never cross the magnetic equator, these drawbacks have little consequence. However, many birds do cross the equator, so it becomes important to know whether they use an inclination compass. Wiltschko and Gwinner (1974) sought to answer this question by performing the same kind of experiment on garden warblers, a species whose summer range includes much of Europe, but which migrates to the tropical and southern regions of Africa. These animals are nocturnal migrants, and Wiltschko and Gwinner found that their orientation bias is preserved when tested in closed chambers in which the stars are not visible. They went on to demonstrate that despite the fact that their migratory route crosses the equator, these birds possess an inclination compass, and they are insensitive to magnetic polarity. Thus the remarkable finding is that all species of migratory birds tested thus far use inclination of the geomagnetic field, even those that migrate between the northern and southern hemispheres.

Despite the apparent disadvantages, these birds have evolved inclination compasses rather than polarity compasses. Most zoologists would agree that this is unlikely to have occurred unless there is a strong advantage of an inclination compass. But what might that advantage be? The most frequently offered answer lies in geophysics. Until the 1960s, geophysicists were content to assume that the orientation of Earth's magnetic field had always been as it is today. After all, in the absence of a good theory of how Earth's magnetic field is generated, there was no compelling reason to consider the possibility that the magnetic dipole could actual perform a flip-flop. However, in the history of our planet, the magnetic poles have actually reversed. In fact, it has happened on a number of occasions. No one knows exactly why or how this takes place. However, that it does take place is no longer questioned. The evidence has been indelibly written in mineral deposits, and geophysists called paleomagnetists have deciphered the coded message.

Paleomagnetists measure the magnetism of rocks. Old rocks. In some cases, really old rocks. Rocks are formed in a variety of ways. Some are produced by accumulation of sediments. Others are produced when molten lava cools. In either case, certain mineral constituents are naturally magnetic, and because of this, they, like the needle of a compass, align themselves with Earth's magnetic field. Paleomagnetists can determine the orientation of the magnetic particles in rocks, so they can determine the polarity of the geomagnetic field at the time the rocks were formed.

It was with a combination of surprise and consternation that paleomagnetists working in France, Japan, and Iceland found that only about half of the lava flows they were studying were magnetized in the northerly direction. The other half were magnetically oriented toward the south! And these lava flows were "only" several million years old, pretty young by geophysical standards. What should a paleomagnetist make of this? Some apparently shrugged it off. But some took their data seriously, and suggested that these reversals in the magnetic dipoles of igneous rocks were produced by reversals in Earth's magnetic field. The scientific reaction was one of, well, silence. However, the examples of reversed magnetic remnants in geological specimens continued to accumulate, so the results could not be ignored.

In the late 1950s, an alternative hypothesis became available. A way out. Under the right conditions, certain minerals can self-reverse. Their magnetic particles can become oriented backward, so that if they are placed in a magnetic field directed north, their particles still orient toward the south. This is a rare effect, occurring in several types of synthetic minerals and a few natural minerals. But it was an alternative explanation to the hypothesis that Earth's magnetic field actually goes through repeated reversals. The self-reversal hypothesis immediately suggested an experiment, however.

What was needed was an organized examination of mineral deposits of known geological age, distributed throughout the world. If all rocks of the same age were magnetized in the same orientation, regardless of their distribution on Earth, this would be strong evidence for the notion that polarity reversals in the minerals were due to polarity reversals of the geomagnetic dipole. However, the data would support the self-reversal hypothesis if it was found that rocks of similar ages, found in different places and formed in different ways, were not all magnetized in the same direction. This is what is known as a critical experiment. A critical experiment is one that, if successfully performed, can have only two possible outcomes. Moreover, each possible outcome is associated with either one of two rival theories. Thus, critical experiments are experiments that lead to the rejection of one theory in favor of an alternative theory. They pretty much can decide which idea is correct, or at least which is a "better approximation of the truth."

The principal problem in performing this particular critical experiment was ensuring that the different samples of rocks, obtained from throughout the world, were actually the same age. The method used to determine the age of the rocks was an atomic clock, the radioactive decay of potassium to the chemically inert gas argon-40. The concept is the same as carbon-14 dating of organic material. Knowing the rate of decay of potassium into argon-40, one can estimate the age of the rock by determining the amount of potassium and argon-40 in the sample. Many samples from lava flows throughout the world were collected. The rocks were not that old, so only trace levels of argon-40 were expected. The dating procedure used equipment that was so scarce, the work was done in only two laboratories, one in the United States and one in Australia. The results from both laboratories were consistent and clear-cut: all the rocks of the same age had the same magnetic polarity, and when the polarity reversed in samples from one part of the world, it reversed in all samples of the same age. Self-reversing mineral deposits were thus demonstrated to be very rare in nature. More important for our purposes, the reversals of the geomagnetic dipole was confirmed.

An explanation for these reversals is beyond the scope of this book, and beyond the scope of this author. It is a fascinating topic, however, and we can refer to several relevant points. First, Earth's magnetic field is thought to be generated by the flow of electrical current within Earth's core. The core consists of molten metals, so it does conduct electricity. We have already mentioned the fact that the flow of electrical current always produces a magnetic field. The mathematical equations that describe the magnetic field produced by current flowing in a circular trajectory are symmetric, in that for each solution that predicts a northerly directed dipole, an alternative solution exists for a southerly directed dipole. Theoretical geophysicists have produced models of magnetic dynamos that will, in fact, reverse their polarity periodically. Many unanswered questions remain, however, including factors that determine the timing of a particular reversal of polarity and factors that determine the average interval between the reversals. Some have suggested that these events are determined by myriad variables, and predicting geomagnetic reversals may be a little like trying to predict when a hurricane will develop. A branch of mathematics called chaos theory, popularized in the motion picture *Jurassic Park,* may ultimately help theoreticians solve these enormously complex problems.

While prediction of geomagnetic reversals is still well beyond the abilities of our most gifted theoretical geophysicists, reading the historical record of the reversals is well within our means. That record shows that polarity reversals have occurred 24 times in the past 5 million years, the last reversal coming about 700,000 years ago. The shortest known polarity interval during the past 4 million years was only 20,000 years long; the longest lasted 700,000 years. That is 24 reversals, at very irregular intervals, since the earliest hominids appeared. And the avian class—the

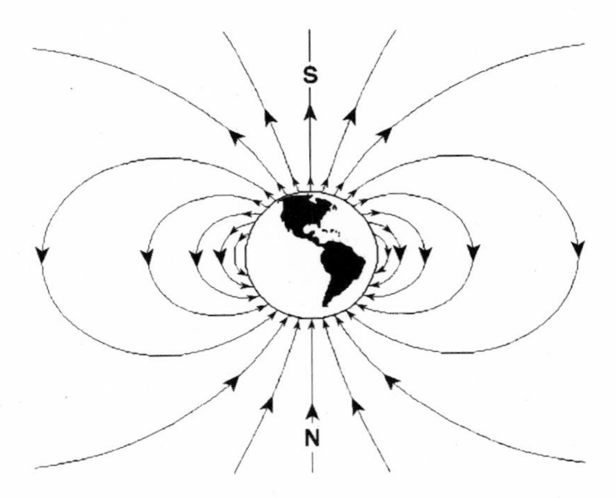

Figure 9.8
The geomagnetic field 700,000 years ago.

birds—have been around much longer than that. Surely they have had to withstand many reversals of the geomagnetic field. (See figure 9.8.)

Consider the consequences of these polarity reversals for a species of migratory bird that evolved a polarity compass. We might suppose that young birds whose summer range lies within the northern hemisphere have an instinctive tendency to orient toward the south each fall. If that instinctive tendency was based on a polarity compass, a reversal of Earth's magnetic field could be catastrophic. However, a sense of magnetoreception that is based only on the angle of inclination would be largely unaffected. Perhaps this is why today's migratory birds possess an inclination compass rather than a polarity compass. Perhaps there were some unfortunate species that evolved a polarity compass, and they are no longer with us.

The Human Compass Revisited

It is interesting (if not completely appropriate at this point) to return our attention to that unfortunate, disoriented denizen of the greatest structures the world has ever known: that peculiar form of mammal that categorizes all other life forms, and in the branch of science known as biological systematics has called itself *Homo sapiens*. The human. Us.

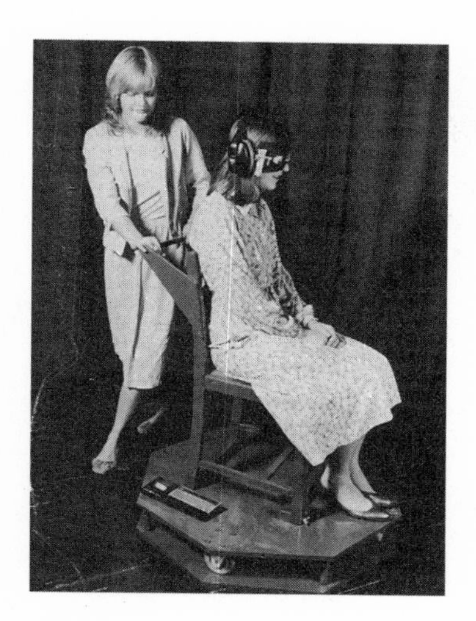

Figure 9.9
Photo of an experiment designed to determine whether humans possess a biomagnetic sense.

Despite our general navigational ineptitude, is it possible that *we* have some form of magnetoreceptive sense? Don't laugh (not just yet, anyway). Some serious scientists think so.

You can probably anticipate at this point how one might go about finding such a thing out. Merkel and Wiltschko have shown us the path; we just need someone to follow it. What if we did the equivalent of a bird orientation experiment with humans? We don't need to put people in a big funnel and spread ink all over the floor. We could simply put someone in a circular room, with no windows, no view of the sky, no obvious navigational cues. For good measure, perhaps we should put the room in a labyrinthine building, so it is hard to keep track of orientation by knowing the path into the room. We could make the person wear a blindfold, and for good measure, let's place the person in a rotating chair and spin it around a few times. Kind of like the children's game Pin the Tail on the Donkey. Now, after all that, could a person—a blindfolded, perhaps slightly dizzy person—point toward the north with any accuracy at all? (See figure 9.9.)

Some investigators have claimed that the answer is "yes" (e.g., Baker, 1985). Oh, sure, the data are noisy, and you still have to perform the second-order statistical analysis we described earlier. But when you do, Baker reports that people score above chance at orienting themselves according to compass directions. Birds do it. Bees do it. Why not us?

Of course, there is this feeling we have that some people are better at it than others (my brother Mark certainly makes this claim), and the research seems to bear that out. In this case, though, when we systematically check a lot of different people and see how well they orient without any obvious cue, we have the ability to try to figure out what *determines* this variance. If some people are better at orienting themselves than others, we can ask why. And we can hope to begin to find some answers. We can sort the people as, say, good, medium, and poor orienters, and then try to find some feature that the good ones have in common and that the poor ones don't have.

Several answers have been forthcoming, and they are worth telling. First, the orientation in which you sleep may be important. Those who sleep with their feet pointing north tend to be the best, followed by those who sleep with their feet pointing south. The north-south orientation tends to produce better results than either east-west orientation. Before jumping to any conclusions, perhaps we should be reminded that many before us have rejected results out of hand and lived to regret it. It might not even be that far-fetched. If there is a magnetoreceptor in the human body, it may perform better when given prolonged exposure to a magnetic field of the proper polarity. So far as I know, no one actually thinks the sleep part of the story is that critical. The assumption is that an appropriate alignment of your body for an extended period of time may be helpful.

Another aspect that has been identified is the type of clothing you wear. Do you like cotton or polyester? Well, maybe no one actually admits to preferring polyester. But it is practical. It doesn't wrinkle, shrink, fade, or stain easily. Maybe parents dress their kids in polyester, even if they don't like wearing it themselves. In any case, polyester is reportedly bad for magnetically based orientation (Baker, 1984). Compare the same people on the same task. In condition 1 they wear cotton; in condition 2 they wear polyester. Performance in cotton is better than in polyester.

Well, polyester is synthetic. It's also a dielectric material, meaning it is an electrical insulator. One way magnetoreception could work is that the magnetic field induces small currents in the body, currents that could activate neurons and produce a sense of magnetic north. The polyester could interfere with that. Stranger things have turned out to be true.

But it does raise an issue. After all, we didn't evolve with clothes, or even with a lot of fur. Maybe all these clothes interfere, to a greater or lesser extent. Maybe many native peoples, who are not obsessed with our Western (especially American?) sense of modesty, have an advantage here. Maybe clothes make the man, but ruin the compass. The idea has been tested. British researchers went to what they delicately call a "naturalist camp," and tested the campers, with and without cotton gowns. Spun them in a chair, naked as the day they were born, and had them point north. Then replicated the whole procedure with the subjects wearing comfortable cotton. The results? Polyester was the worst, but cotton was no different from "nothing at all." Well, for some that is a relief; for some, perhaps, a disappointment. In any case, it is something to think about. For example, much of our space age winter clothing—skiing gear, that sort of thing—is made of synthetic material. Something to think about when you're cross-country skiing in the backcountry, miles from anywhere. Just make sure you take your compass, because if you get lost, some of the alternative means of navigation could leave you mighty cold!

Before we leave this topic, it is important to mention that attempts to replicate these orientation effects in humans have been tried, and the majority have failed to replicate Baker's findings. Many experts feel that humans do not have any magnetoreception capabilities, no matter what they wear. So I wouldn't want to mislead anyone—unlike birds, the possibility of magnetoreception in humans has not been proved to the satisfaction of most experts.

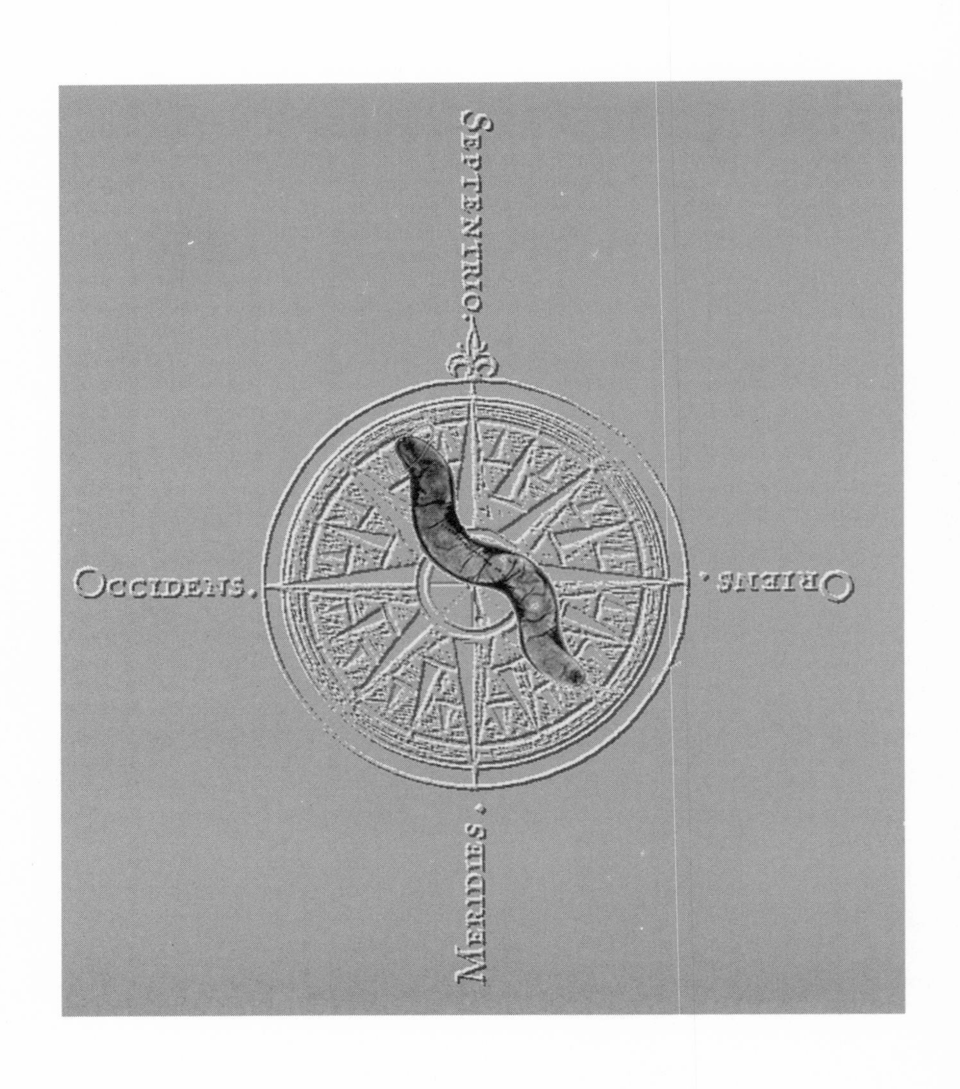
Septentrio.
Oriens.
Meridies.
Occidens.

10

The Search for the Magnetoreceptor

Little Magnets in the Head?

Well, this is all very fascinating. But when you are trying to establish the existence of a new sensory modality, you really need a receptor. It's like a habeas corpus. The authorities are not going to completely accept the evidence of a new sensory modality without identification of the receptor. It's not an unreasonable requirement, although the ultimate evidence for a sensory modality in any living creature has to be evidence that the creature actually *uses* it. For example, dogs and cats clearly have the types of photoreceptors that are required to see in color. They *have* the receptors, but do they *use* the receptors? No, they really don't. They are essentially color-blind. So having the receptor is not necessarily enough. Just the same, when you have evidence that a certain kind of environmental energy is used by an animal (and we do seem to have that sort of evidence for magnetoreception), it becomes important to find the receptor that transduces that energy into neural signals. And herein lies the major problem in magnetoreception. Despite years of trying, we have yet to identify a magnetoreceptor. Often, however, the quest for a solution is almost as interesting as the solution itself, so let's spend a few pages finding out what has been done until now.

The first question we must formulate is how a magnetoreceptor might operate. The most obvious answer comes from a compass. Perhaps migratory birds actually possess a compass, a little piece of magnetic material that responds to the lines of force in the terrestrial magnetic field. But before considering the possibility that magnetoreception relies on receptors composed of little magnets, perhaps we should describe magnetic substances in general a little more.

The naturally occurring magnetic substances are referred to as "ferromagnetic." As it happens, iron and certain other elements, including cobalt and nickel, are strongly attracted by a magnetic field. Other substances, such as aluminum and platinum, are only weakly attracted by magnetic fields. These are called paramagnetic substances. Finally, some matter is repelled by a magnetic field. Copper and silver are examples of what are termed diamagnetic substances. What determines whether a particular form of matter will be ferromagnetic, paramagnetic, or diamagnetic? These properties are determined by the behavior of electrons within different kinds of atoms.

We are all familiar with the idea that electrons move in orbits around the nuclei of atoms. As they move in their orbits, however, they also spin around an axis of rotation. The motion of electrons is not unlike the motion of Earth around the sun: not only does Earth rotate around the sun (once each year), it also rotates on its axis (once each day). Electrons also spin as they circle the nucleus. The thing that is unique about ferromagnetic substances is that the electrons can align the axes of their spin, to be in the same direction. Each electron's spin gives rise to a small magnetic force called a magnetic moment. When many electrons share the same axis of spin, their individual magnetic moments reinforce each other, and the atom as a whole displays magnetic properties. The atoms of paramagnetic substances also have magnetic moments, but their alignments tend to be random. The atoms of diamagnetic materials tend to be made up of atoms that have an even number of electrons, and are therefore magnetically neutral. When they are placed within a magnetic field, however, diamagnetic atoms will orient themselves in such a way that they are repelled by the field.

So ferromagnetism depends upon the behavior of electrons within the atoms that make up matter. The atoms of ferromagnetic materials are naturally aligned so that clusters of atoms each have a magnetic moment. However, while all minerals containing iron are attracted by a magnet, not all pieces of iron act like magnets. The explanation for this paradox is found in what are termed "magnetic domains." A magnetic domain is a collection of molecules whose magnetic moments are perfectly aligned (all north poles pointing in the same direction). This small portion of the material is magnetic. However, in the unmagnetized state, the material is made up of many different magnetic domains, and each points in a different direction. Thus, larger aggregations of the material are not magnetic. Application of a strong external magnetic field can align all the different magnetic domains, and thus create an artificial magnet.

Some ferromagnetic materials are natural magnets. The most familiar example is the mineral magnetite, which is a common ore of iron and has the chemical formula Fe_3O_4. It is also known as lodestone. Pieces of lodestone attract other pieces of lodestone, and they also attract metallic iron. If a thin piece of lodestone is hung by a thread, it aligns itself with the lines of force in Earth's magnetic field. It points toward the magnetic poles, and if it is small enough and properly balanced, it also aligns itself with the angle of inclination of the geomagnetic field.

Now if we took a needle made from lodestone and cut it in half, each half would act like a magnet. If we cut the halves in half, they would also act like magnets. We could go on with this process until we had a tiny piece of lodestone that was composed of a single magnetic domain: all of the magnetic moments would point in the same direction. There would be no variability. This would be a small magnet, but one powerful for its size. Such a uniform piece of magnetite is called "single-domain magnetite."

What if magnetoreceptors were endowed with microscopic pieces of single-domain lodestone? They would then attempt to align themselves with Earth's magnetic field. Their attempts to orient along the lines of magnetic force could then be detected by a special type of sensory receptor called a mechanoreceptor. Mechanoreceptors are commonly found throughout the body. Our skin contains thousands of them. When they are subjected to a mechanical force (light touch or pressure), their nerve membranes are deformed, and this allows ions to flow between the inside and the outside of the nerve cell. Since ions are electrically charged particles, such movement is a form of electrical current known as a

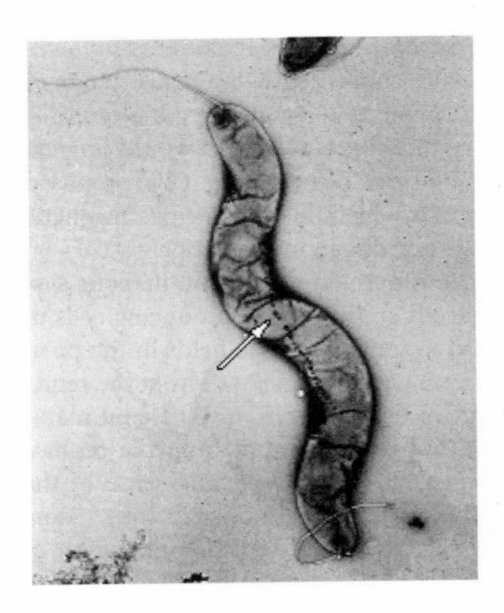

Figure 10.1
Electron micrograph of a marine bacterium that possesses a true magnet. In this illustration, single-domain particles of magnetite appear as a small chain of black spots (see arrow).

bioelectric current. This current leads to nerve impulses in sensory nerve fibers. Voila, we have a magnetoreceptor!

At first, the idea seemed preposterous. How could biological organisms manufacture or synthesize magnetic material? Could they simply incorporate it from the environment? Maybe through their diet or something? Then, in the midst of all this skepticism, an organism was found that is indeed endowed with single-domain particles of lodestone. It wasn't a bird, or even an insect. It was a lowly form of bacteria. Bacteria that contain little rods of single-domain lodestone. And sure enough, these little marine bacteria respond to Earth's magnetic field. Not in any way that we would wish to call navigation: their magnetite particles simply serve to align the bacteria with the lines of geomagnetic force. This is a completely passive process, but it ensures that these magnetic bacteria will point along the lines of inclination. These bacteria live in the sediments at the bottoms of lake beds and the ocean. They are anaerobic, and oxygen will kill them, so anytime water turbulence stirs the bottom and brings the bacteria out of their oxygen-poor habitat, they must swim back toward the sea bottom. The magnetosomes located within these single-cell organisms allow them to find the seafloor. (See figure 10.1.)

The way it works is pretty interesting. In the northern hemisphere, most of the so-called magnetotaxic bacteria have the the north-seeking pole of their magnetosomes facing frontward, and they have a little tail (called a flagellum) in the back. Just like little magnets, the strength of their magnetosomes is sufficient to ensure that their front end will always face north. Thus, all they have to do is swim forward, and the torques produced by the interaction between their magnotsomes and the geomagnetic field ensure that they will always swim along the lines of magnetic force, which for north-seeking bacteria located in the northern hemisphere is always toward the sea bottom, along the lines of magnetic inclination. Pretty clever.

Based on these observations, we should not be surprised to find that magnetotaxic bacteria in the southern hemisphere have the *south-seeking* poles of their magnetosomes facing forward, so they will swim toward the south magnetic pole. Once again, the lines of inclination ensure that they will reach the bottom. What happens at the magnetic equator, where the lines of magnetic force have no inclination (they are parallel to Earth's surface)? Well, the north-seeking and south-seeking bacterial are found in equal numbers. What is more interesting is that not all magnetic bacteria found in the northern hemisphere are of the north-seeking variety. A certain proportion of south-seeking bacteria are always found in northern latitudes. Similarly, a small (but significant) proportion of north-seeking bacteria are found in the southern hemisphere. These "wrongheaded" bacteria are definitely at a competitive disadvantage, since their magnetic compass will lead them away from the seafloor rather than toward it. So why hasn't evolution eliminated these "wrong way" bacteria?

The answer may lie in the reversals of the geomagnetic field. North-seeking bacteria living in the northern hemisphere *currently* have an advantage because the northern magnetic pole is *currently* in the northern hemisphere. But we know that has not always been the case, and we fully expect another reversal in Earth's magnetic dipoles. In fact, we might consider that the next reversal is already a little late! In any case, if all the bacteria were the same, then the whole species might become extinct when the next reversal does occur. However, maintaining a small population of "backwards" bacteria in each hemisphere ensures that there will always be a population that will be favored by successive reversals in the geomagnetic field. Even more clever, don't you think?

Well, magnetotaxis in bacteria is interesting, but it's not exactly what we were looking for. The size of the single-domain magnetosomes is just large enough to produce magnetic moments that will orient these small, one-cell organisms. But this is not the same as providing magnetic receptors in a large bird, a tortoise, or even an insect. The discovery of these little magnetic bacteria was a start, however. It at least established the existence of single-domain magnetite in one form of life. It certainly seemed possible that magnetite may be present in other species and that it could play a role in magnetorception. The discovery of magnetotaxic bacteria thus provided the incentive for an intense search for magnetite deposits in magnetically sensitive birds and insects.

The search for ferromagnetic material in larger animals has been in progress for some 20 years now, and it has not gone entirely unrewarded. Single-domain magnetite has been found in many migratory birds, in homing pigeons, in sea turtles, in honeybees, and yes, even in humans. It is worth noting that the bodies of animals are completely transparent to magnetic fields (if that were not the case, magnetic resonance images of the inside of the body would not be possible). This means that the magnetoreceptor would work no matter where in the body it was located—the search has focused in the head, but the magnetoreceptor could be *anywhere*.

Why, then, the emphasis on the head? Perhaps because earlier work had shown that small magnets placed around the head could disrupt homing in pigeons. What appear to be single-domain magnetite particles have been found on the inside surface of the skull in birds, rodents, and humans. Interestingly, magnetite deposits are often found around the nasal cavity, and although there is little or no direct evidence, it has been suggested that magnetoreception and the sense of smell may be related in some way. The association between the nose and magnetic sensitivity was apparently made by the early Mayan peoples of Central America. Professor Vincent Malmström, a colleague of mine in the Department of Geography at Dartmouth College, made a mysterious discovery while working at Izapa, an ancient mesoamerican ceremonial site in Mexico (Malmström, 1976). Professor Malmström discovered that a basalt sculpture of a turtle was carved in such a way that the head was magnetic, and all the lines of magnetic force were focused on the animal's snout!

While magnetite has been found in tissues of many animals, that alone does not constitute evidence that magnetite serves as a magnetoreceptor. In order to serve this function, the magnetite must respond in some way to Earth's magnetic field. Just as important, information concerning these responses must be conveyed to the central nervous system. In terms of neurobiology, we would say that in order to qualify as a potential magnetoreceptor, the magnetite particles must receive an innervation. And herein lies a major concern: no one has been able to show that the magnetite that has been found has any contact with nerve cells. At least that is the case in vertebrates. Magnetosomes that do appear to receive an innervation have recently been described in honeybees.

That pretty much summarizes the state of affairs with respect to a ferromagnetic basis for magnetic perception—at least it did until late in 1997, when Michael M. Walker and colleagues at the University of Auckland published their findings on the mechanism of magnetic sensitivity in trout. Walker et al. (1997) describe a beautiful series of experiments that enabled them to localize a magnetoreceptor for the first time. Their experiments are concise and elegant. They include virtually all the elements that we regard as essential criteria for identification of a sensory modality and the associated receptor mechanism. First, Walker et al. showed via behavioral testing that the trout is sensitive to perturbations in Earth's magnetic field. They did this by training the fish to strike at a target that was distinguished by a focal change in the intensity of the magnetic field. Since the fish could learn to strike at such targets and avoid alternate targets that were only distinguishable by the magnetic cues, the behavioral experiments confirmed that the trout can sense the magnetic field. Since magnetite had been previously found in trout's skulls, the investigators decided to record neural activity from nerves that innervate the relevant region of the skull. They discovered a population of nerve fibers that respond to changes in the ambient magnetic field in one specific nerve, called the ros V ("ros five" nerve). This is a branch of the trigemminal nerve, which supplies an innervation to the face and skull of all vertebrates, including humans. Not all the nerve fibers in the ros V nerve respond to magnetic stimuli, just a specific subgroup. The important point, however, is that magnetically sensitive fibers were identified.

The next step was to search for an anatomical structure that contained magnetite. Using a combination of light and electron microscopic techniques, Walker et al. found magnetite deposits in the trout's nose. Specifically, the magnetite was found in a layer of tissue that was directly beneath the olfactory receptors themselves. The magnetite particles are very similar to those found in magnetotaxic bacteria. The big question now was, could they demonstrate that this tissue was innervated by magnetic-sensitive fibers in the ros V nerve?

To answer this last and most critical question, Walker and co-workers injected minute quantities of a special dye in the ros V nerve. The injections were made into regions of the nerve that contained magnetically responsive fibers. The dye migrates within the axons of nerve fibers and can be visualized using appropriate microscopic techniques. If you wait long enough, the dye will migrate all the way to the beginning branches of the nerve. So, if these color-coded fibers could be seen terminating around the magnetite-containing cells in the olfactory lamellae within the trout's nasal cavity, then the first magnetite-based magnetoreceptor will have been discovered. And that is exactly what was observed—fibers containing the dye were found to ramify all around the magnitite-containing cells within the olfactory tissue. Figure 10.2 illustrates the magneto-sensitive system of the trout as revealed by Walker et al.

Of course, many issues are not yet resolved. It appears as though the magnetic sense of the trout is not sensitive to the polarity of the field, but it's not yet clear that the trout has an inclination compass either. The behavioral results of Walker et al. clearly demonstrate the trout's magnetic sensitivity—they do not demonstrate the sort of inclination compass like that possessed by migratory birds. More generally, it's not at all clear how any ferromagnetic receptor could endow its owner with an inclination compass without also providing a polarity compass. It is also generally accepted that the sensitivity displayed by birds to experimental manipulations of the ambient magnetic field are only effective within a range of field strengths that approximated the earth's magnetic field. Birds are largely insensitive to the application of magnetic fields that are much weaker or (perhaps more suprising) much stronger than the earth's

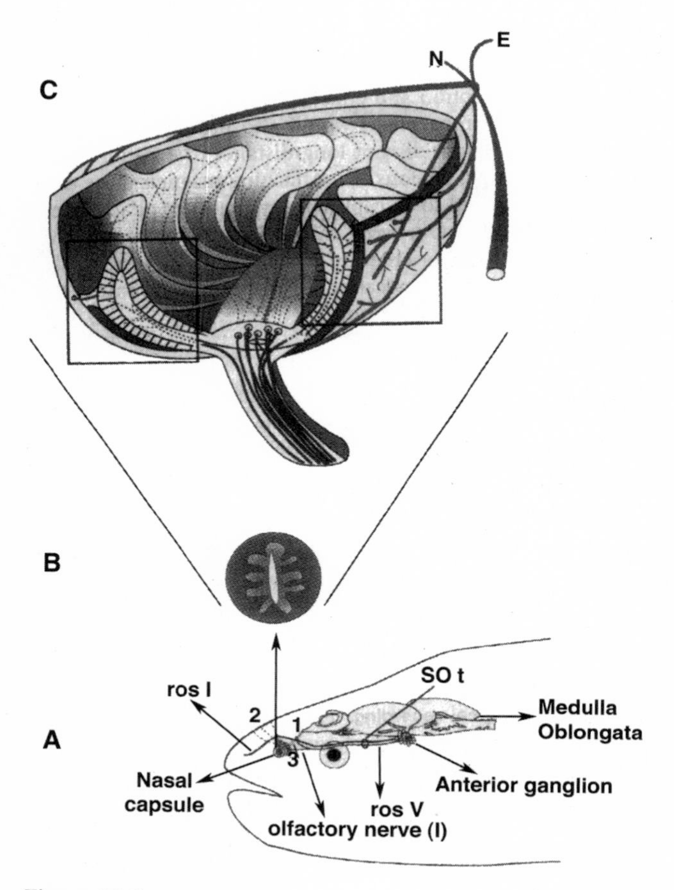

Figure 10.2
Illustration of the location and morphology of the putative magnetoreceptor of the trout. The olfactory nerve innervates the olfactory receptors and carries information related to the sense of smell. The ros V has a branch that innervates the magnetoreceptors, which, along with olfactory receptors, are located in rosettes (3).

magnetic field. Unlike the birds, the effects of a magnetic field on magnetite would only be expected to increase with increasing field strength. Yet, the results of Walker et al. provide the first evidence for a ferromagnetic receptor in a vertebrate. Others have wondered whether some entirely different mechanism might be at work, at least in birds and some other species. But what alternative mechanims might there be?

There are two alternative ways that magnetoreceptors could operate that have received serious consideration. One involves the relationship between magnetism and electricity, and the other involves, of all things, vision.

The electrical hypothesis is interesting, but it could work only in marine habitats. We'll have more to say about senses involving electricity and magnetism later on. For now, we can simply note that the possibility of visually mediated magnetoreception appears viable for adaptation by land animals.

A Compass in the Eye?

In 1977, M. J. R. Leask published an article entitled "A Physico-chemical Mechanism for Magnetic Field Detection by Migratory Birds and Homing Pigeons" in the prestigious scientific journal *Nature*. In that short (two pages) article, he pointed to the problems encountered by the ferromagnetic hypothesis and suggested an alternative based on the idea that the pigment molecules of photoreceptor cells in the retina could display a kind of magnetic sensitivity that depends on their responses to light.

The sense of vision begins when a packet of light called a photon is absorbed by a photosensitive pigment molecule. Hundreds of thousands of these molecules are in each photoreceptor cell within the eye, and the eye contains millions of these photoreceptors. Color vision depends on having more than one type of pigment molecule, but each individual receptor cell contains pigment molecules of only one type. When a pigment molecule absorbs a photon, it changes its state. This change ultimately gets translated into minute electrical currents, which represent the beginning of neural processes that underlie vision.

Leask suggested a way that the photopigments might provide information about the orientation of Earth's magnetic field. He proposed that upon absorption of a photon, pigment molecules might be forced into what are normally very rare states in which the electrons in the molecule interact with magnetic fields. While these states might be very rarely attained, continued exposure to light might maintain a certain number of the molecules in the excited state. This process is known as optical pumping, so the general theory is sometimes called the optical pumping hypothesis. There are actually three possible variants of these excited states, and the relative proportions in which they are found are sensitive to external magnetic fields. Eventually, each of these molecules will fall back into a more stable ground state, and when that happens, a photon is released. An important feature of the emitted photon differs according to which state produced it, so the essence of the proposal is that the bird's retina might be able to provide information concerning the surrounding geomagnetic field by analyzing the characteristics of the light produced by pigment molecules as they decay back to their ground state.

Leask's hypothesis was offered primarily to encourage new conceptual approaches to studies of the magnetoreceptor. Although it is based on the quantum mechanics of large molecules, the ways in which a biological system might actually implement the hypothesis were purposely lacking in detail, and Leask himself noted that the hypothesis may or may not be feasible. It did raise some interesting points, however. First, Leask suggested that the low probability that visual pigment molecules could attain one of the excited triplet states fits with the repeated finding that magnetically based orientation is not spectacularly precise. Second, the proposal clearly requires that this form of magnetoreception depend on the presence of light, which of course is needed to achieve the optical pumping of molecules into the excited triplet state. Leask noted that birds are sensitive to only a small range of field strengths that approximate the strength of the geomagnetic field. Experiments in which the field strength is substantially smaller than or *greater* than Earth's magnetic field appear to disorient the birds. Leask pointed out that it was hard to see how a receptor based on ferromagnetism would require such a narrow range of field strengths, but that the effects of magnetic fields on chemical reactions are often dependent on field strength as well as the orientation of the field relative to the molecule (which an inclination compass seemingly requires).

There are certainly problematic aspects of the hypothesis. Perhaps the most important difficulty concerns whether or not optical pumping could produce emitted photons in sufficient numbers to permit the sort of analysis that would be needed in order to provide information about the magnetic field. A second, related problem concerns how a visual system could distinguish between the photons emitted as a result of optical pumping and those that entered the eye and were absorbed during the normal process of vision. To many, these problems seemed insurmountable, and for at least 10 years Leask's optical pumping model was largely ignored in favor of ferromagnetic-based models.

Some recent work has led to a resurgence of interest in the role of the retina in magnetoreception. Although there is not yet direct support for optical pumping, there are some tantalizing hints that the retina and a related structure called the *pineal body* may be involved. To some extent, these new developments were spurred by frustration over all this magnetite business. The search for a (or for that matter, any) magnetoreceptor led some scientists to seek alternatives to a "frontal assult" on the magnetoreceptor. What are the alternatives? One might be to temporarily suspend the search for the receptor, and to search instead for neurons that respond in some way to changes in the ambient magnetic field.

That approach has its own problems. First and most important, finding a population of neurons that responds to magnetic stimulation does not necessarily help identify the receptor mechanism. Logically, one might think that if you found a population of neurons that respond to magnetic stimulation, you could in principle work backwards and find the source of those responses. Practically speaking, that is often not possible. Consider, for instance, that in a "typical" mammalian nervous system, each neuron may have input connections with as many as 1,000 other neurons. Each of these 1,000 input neurons may receive inputs from 1,000 additional neurons, and so on. When you consider that these input cells may be widely distributed within the central nervous system, it is easy to see how difficult it might be to start somewhere in the middle and try to trace the path backward to the ultimate source of magnetic sensitivity. A second obstacle is How do you know where to look? Even a small brain gets pretty large when you study it one nerve cell at a time.

However, Leask's article did serve to focus initial attention on the retina, in addition to brain structures that are known to receive retinal inputs. The first experiment to successfully record changes in the activity of single nerve cells in response to magnetic stimulation was performed by Peter Semm at the University of Frankfurt in Germany. Semm published his findings, some of which were obtained in collaboration with Wiltschko, in the early 1980s. Where were these magnetically sensitive cells? Well, they weren't in the retina, but in an interesting brain structure called the pineal body.

The pineal body is interesting for a variety of reasons. For one thing, it has attracted the interest of anatomists since classical Roman times. These early neuroanatomists noted that the pineal body was the only unpaired structure in the human brain. That is, the brain is bilaterally symmetric, so, to a first approximation at least, every structure found on the right side has a counterpart on the left side. The pineal body, however, sits in the middle of the brain, conspicuous in its singularity.

The uniqueness of the pineal body led the great French mathematician-philosopher René Descartes (1596–1650) to attribute to it a special role in his theory of the relationship between the mind and the body (a question many of us still grapple with today!). Descartes believed that none of our beliefs should be accepted without rational justification, and he argued that the ultimate validity of such justifications is based on mathematics. He claimed to have stepped outside of all of his own beliefs, and to have examined in a rational way whether each one could be justified. One of the thorniest of his metaphysical questions was whether he himself actually existed—or was it possible that he was some devilish illusion or hallucination? Descartes decided that even the most insidious illusion could not produce a mind that could question its own existence, and on that basis decided "Cogito, ergo sum" (I think, therefore I am). Descartes's approach to philosophical and scientific questions was called rationalism and, being based on mathematical principles, was supposed to be incapable of producing erroneous conclusions. It is thus doubly ironic that his methods of rational analysis often led Descartes astray.

We've seen how it is a characteristic of scientific ideas that they do not have to be correct in order to be useful and, indeed, further our understanding. The ideas just have to concern themselves with important problems, stir the imaginations of others, and invite tests of their accuracy. Certainly this was true of Descartes's erroneous ideas. For example, he thought that movements were produced when animal spirits flowed into muscles. When this happened, the muscles would become extended (flex your bicep), and movements would result. The idea was tested in an ingenious way. The logic was as follows: if Decartes's theory of mucle contractions was correct, the volume of a muscle would increase when the muscle was flexed. This prediction was tested by immersing an arm in water and carefully monitoring the level of the water as the volunteer subjects flexed and relaxed their arms. The level of water did not change, so Descartes's theory of muscle contraction was disproved.

Descartes was what philosophers call a dualist: he believed that the mind and(or) soul are different (and separate) from the body. Descartes's rationalist approach led him to conclude that God existed, and that God had created two types of "substance" in the universe: a thinking substance that he called "mind," and an extended substance that can be characterized by geometry, which he called "body." Although he believed that minds and bodies are separate, Descartes suggested that they interact through the pineal body. (See figure 10.3.) In Descartes's theory of life and cognitive science, the pineal body lay at the interface between mind and body. There it "reflected" animal spirits between different parts of the brain and between the brain and the muscles. This idea of the pineal as a reflecting device apparently grew from Descartes's scientific observations on the interactions between light and matter, a discipline now known as optics. A true giant in the history of science, Descartes discovered the law of reflection, one of the principal laws of what we now call geometrical optics. The law of reflection states that when light is reflected off a surface, the angle of reflection equals the angle of incidence (rather like a bank shot in billiards). He thought that the pineal body was like a mirror that reflected animals' spirits and thus caused "animation" (movements). It was the mind that controlled the position of the "mirror," and, thus, voluntary movements.

With respect to Descartes's views on God and mind, and their relationship with the physical world, science has little to contribute. So far

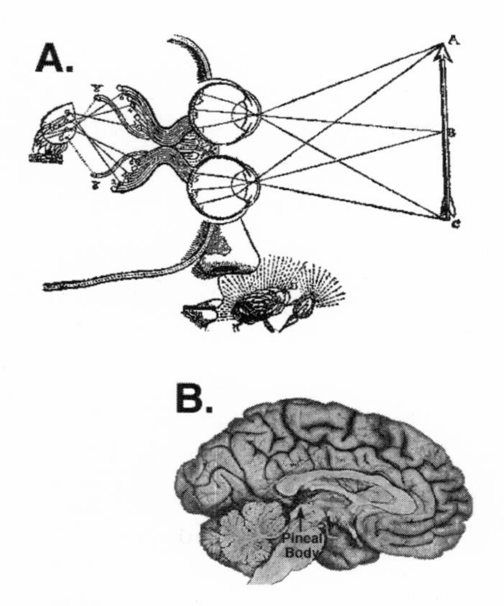

Figure 10.3
The great French philosopher/mathematician/scientist René Descartes thought the pineal body was the interface between the separate spheres of mind and body. Part A illustrates his theory of how sensory inputs converge onto the pineal body. The inputs depicted in this case come from the eyes and the nose. In B, we see the actual position and size of the pineal body in the human brain.

as we know, however, Descartes's theory of the function of the pineal body is completely incorrect. What does modern neuroscience have to say about the pineal?

First, it has the characteristics of both a neural structure and an endocrine gland. Endocrine glands secrete biologically active substances into the bloodstream. Since every cell in the body receives a blood supply, they are all exposed to the secretions of endocrine glands, which are called hormones. Obviously, endocrine glands are very important, because they have the power to influence every cell in the body. Endocrine glands are not autonomous, however. In fact, they operate in close association with the brain. Not only can the brain control the endocrine system, but the endocrine system can in turn control the brain. But right now, let us return to the object of Decartes's fascination, the pineal body.

We've just pointed out that the pineal body has some glandlike properties. For one thing, it produces a hormone called melatonin. One interesting feature of melatonin is that the levels of this hormone vary throughout the course of the day. Melatatonin levels fall during the day and rise again each night. This rhythmic pattern of changing melatonin levels is an example of what is called a circadian (circa = about; dian = day) rhythm. These rhythms have a period approximately a day in length, and are but one manifestation of what are called "biological clocks." Many bodily functions have cycles that are approximately one day in length. Body temperature, our mental acuity, and our sleep cycles all have daily rhythms. Other body functions are periodic, but the periods are much longer than a day. The menstrual cycle is one that immediately comes to mind, but there are also seasonal changes, like animal migrations. Clock-type metaphors are often invoked when dicussing the aging process, and ultimately we refer to life as a cycle. We don't want to make the same mistake Decartes did: many of life's cycles probably have little to do with the pineal. The main point we wish to make is that the pineal body appears to be intimately related to biological clocks: at least it appears to be essential to a variety of circadian rhythms.

They are called *circadian* rhythms because they are approximately 24 hours in duration. For example, our natural sleep cycles last a little longer than 24 hours. This has been repeatedly demonstrated in experiments in which human volunteers are isolated for long periods of time. A classic example is the experiments in which human volunteers decended into a limestone cave in Kentucky. They were provided with any reading materials they wanted, but access to other media, such as TV and radio, was strictly controlled. The reason? These people were to remain completely unaware of the time of day aboveground. The had what many of us may think of as the ultimate lifestyle: they could go to sleep when they were tired, wake up when they felt well rested, nap as they saw fit, and work only when they felt like it!

What invariably happens in this situation is that people stay up a little later each night and wake up a little later each day—their "natural" activity cycle is a little more than 24 hours. Eventually, they are staying up all night and sleeping all day, blissfully unaware that they are completely out of phase with the rest of the world.

The question therefore arises: Why is it so hard to work the night shift? Why do we have all this concern about human performance when people stay up all night? The reason is that the *light cycle of the day synchronizes these biological clocks.* They are entrained by light. It is only when we are completely isolated from Earth's day-night cycle that our biological clocks fall into the "free running" mode and show what their endogenous period really is (approximately 25 hours). Light keeps us "in time" with the planet.

How do we sense that light? Well, most animals use their eyes. And for most species, the synchronization of circadian rthythms depends on vision. But birds are curious, and one of their curious features is that their pineal bodies respond to light—directly, through their skulls!

Maybe a conceptual link is developing here. There are animals that can demonstrate a certain sensitivity to Earth's magnetic field. The biases in their premigratory hopping demonstrate their sensitivity to the inclination of the lines of force; the angle between the magnetic force and the force of gravity. What kind of sense organ do they possess? Ferromagnetic receptors are possible, and the animals have small deposits of magnetically stable magnetite in their bodies—often in their heads, and often in close association with the nasal cavities. We can't establish any direct connection between these magnetite deposits and the nervous system, however, and that's a problem. So, in a search for an alternative form of receptor cell, Leask points out that the quantum mechanics of macromolecular chemical reactions can be sensitive to the orientation of a magnetic field, and that this may form the basis of a sense of magnetoreception. In a parallel development, the responses of brain cells to changes in the orientation of the magnetic field indicate that certain nerve cells are indeed sensitive to changes in the ambient magnetic field. Where are those cells? Well, at least some of them are in the pineal body. And what completes the circle? Well, it turns out that the pineal bodies of birds are unique—they respond directly to light that diffuses through the skull. Thus, we have a link between light and magnetoreception! What is more, we have a link between magnetoreception and migration, since one major cue that the migratory season is approaching is the length of the daylight cycle.

Devilishly intricate, yet exquisite. And to add fresh tracks to the trail, we discover that the pineal body and the retina share remarkably similar biochemical features, suggesting that the retina may also display a magnetically induced vartiation in visual responsiveness. Yes, it does happen.

Responses of retinal cells do vary with magnetic field orientation, and so do certain populations of cells within the central nervous system proper that receive direct inputs from the retina.

This is an intriguing tale of circumstance, veiled truth, and innuendo. Implications and inferences abound. But it may yet be a case of mistaken identity and misinterpretation of the clues. We'll have to see. We still don't know how these magnetically induced responses are actually transcribed into nervous activity, which, after all, is what we set out in search of. But you do get the feeling that we are much closer to solving one of the important parts of the remarkable mystery of animal migration. We are indeed making progress on a question posed by Middendorf 140 years ago (and perhaps by the ancient Mayans centuries before that): Do animals have a magnetic compass?

11

The Sun Compass of Bees and Ants

Can you tell where the sun is in figure 11.1? I mean, can you tell where in the sky the sun is, simply by looking at an expanse of pure blue sky? Or how about this? Imagine you are lost in a dense tropical forest like that pictured in figure 11.2. Do you think you could determine the position of the sun by looking at the several small patches of sky that are visible through the dense forest canopy?

It seems absurd to ask whether you can tell where the sun is without actually seeing it, and for us, it is. We really can't do it. Maybe we could get a rough estimate by examining the shadows, but how much confidence would you place in the estimate? Enough to determine which way is north? Almost certainly the answer is an emphatic no. But such navigational feats are no challenge at all for your ordinary, garden variety honeybee or, say, the ants that inhabit the Sahara Desert. So you wonder why a little insect should need to be able to determine the position of the sun from a small patch of blue sky. Of what use could that be? Why bother?

Well, consider first the fact that bees and ants are social insects. Indeed, they have some of the most complex societies in the animal kingdom. Each colony can contain hundreds of thousands of individuals. Such a large community has needs. Needs for shelter, for food, for taking care of offspring. It's a big job, and it takes a highly organized society to do it. Indeed, it is often said that the whole insect colony acts like a single organism, and if that is so, then the individual bees or ants are like individual "cells" that together comprise this enormous "metaorganism."

Figure 11.1
The magnificence of the Presidential Range of New Hampshire on a clear day in October.

In any case, these enormous insect colonies have to eat. If an animal in the wild wants to eat, it has two choices. It can sit still and wait for something that looks like, sounds like, smells like, has an electric field like, or has the temperature of food. Or it can take the initiative and go out there and find food. In the branch of science that deals with animal behavior in the wild, the search for food is called *foraging*. I'm not sure if anyone even has a name for the former strategy. *Ambush* seems appropriate, but it doesn't sound scientific. *Quietly lying in wait* is an accurate enough expression, but it isn't very catchy.

In any case, bees and ants can't just wait for the food to come to them. Not many flowers will just wander past the old beehive! No, bees have to go to the food. They have to forage. And when they find a nice rhododendron, they are going to have to let all the other members of the colony know exactly where this wonderful new source of food is. The needs of the colony are such that many workers must help collect the nectar. Foraging requires an ability to find a food source, return to the

Figure 11.2
The Tijuca Forest, Rio de Janeiro, Brazil.

hive, and convey the location of the source to other members of the colony so they can help collect the nectar. All that requires navigational skills, and that's why we're bringing it up.

The scientific study of insect navigation can fairly said to have been started by the work of Karl von Frisch. That may sound an esoteric accomplishment, but Frisch was awarded a Nobel Prize in medicine or physiology in 1973. Many people have heard of the bee's famous dance, the "waggle dance"—a behavioral form of honeybee communication discovered by Frish. It's been on TV animal shows, is featured in many textbooks on animal psychology and zoology, and has been in the popular press. You may even have heard people say that Karl von Frisch won the Nobel Prize for studying a "dance performed by bees." I don't think that is why he won the Nobel Prize. I think he won it because of his contributions to understanding bee navigation. But those contributions depended upon Frisch's deciphering of the messages conveyed by the bees' waggle dance. So even though it may at first sound esoteric, maybe

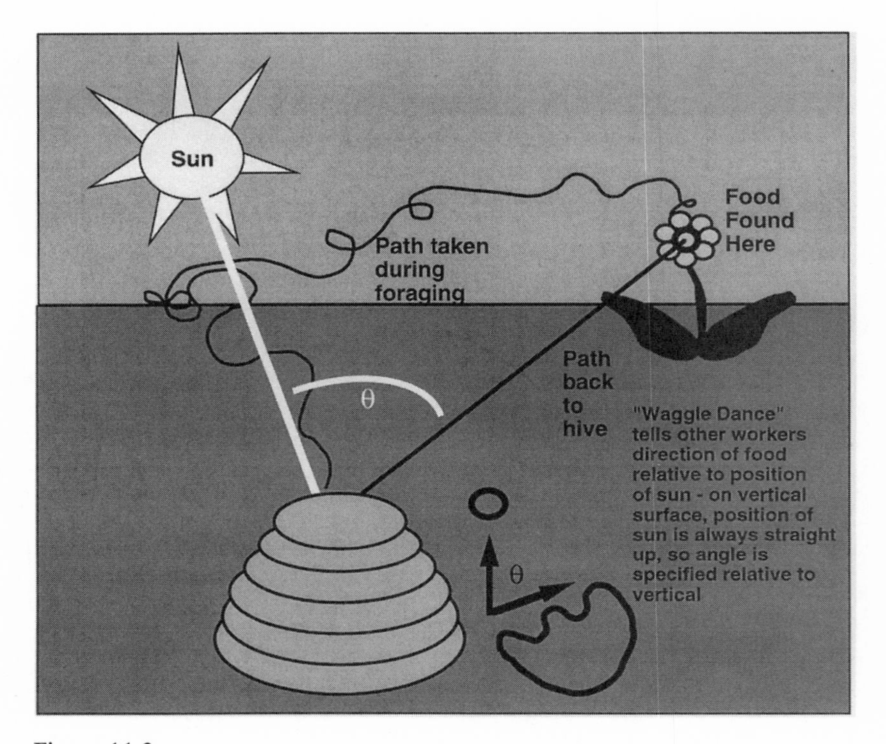

Figure 11.3
The navigational abilities of honeybees are illustrated in this figure. Despite the circuitous route a bee might take before finding food, the return trip is always quite direct. Upon returning to the hive, the bee performs the "waggle dance," which communicates the location of the food source to other bees. Research has shown that two important types of location information are represented by the dance. The first is the distance of the source from the hive, and the second is the direction of the source relative to the location of the sun.

we should spend a little time describing the scientific accomplishments of Karl von Frisch.

When worker bees leave the hive, they usually are searching for food. They may wander over half a mile or more in search of nectar: an impressive distance when you consider their small size. As they search for food, they meander here and there, so their paths can be quite complicated (figure 11.3). When they finally find a good source, they head directly back to the hive (the likely origin of the expression "make a beeline"?). So we pose a question: How can the bees, after all this wandering, "know" exactly which direction will take them directly back to the hive? Desert ants find themselves in a similar situation, which is perhaps made even more difficult by the lack of distinctive landmarks that typically characterizes desert landscapes.

This example reveals some of the navigational problems encountered by foraging insects. Their travels may not span thousands of miles, like those of migratory birds, but the need for navigational cues is still important. Like the birds, bees can use a variety of different cues. This does not surprise us, for we've seen before that something as important to an individual as navigation is not likely to rely on one cue alone. Almost all of the navigational cues used by honeybees were initially discovered by Karl von Frisch, his students, and his coworkers.

Early in his work, Frisch found that one of the navigational aids used by bees is a sun compass. That shouldn't surprise us—the sun is a conspicuous feature of the world, and it can provide information useful for navigation. But what might be surprising is the elaborate "language" that bees use to communicate the location of food sources to the other workers in the hive.

When a bee finds a good food source, it returns directly to the hive, and communicates the location of the source by using an elaborate dance, called the "waggle dance." (See figure 11.3.) The dance steps go something like this. The bee walks a short distance, wiggling its body as it goes. The quality of the source is signified by the vigor of the wiggling motion, and the distance is indicated by the length of the walk. At the end of the wiggling phase, the bee returns to the starting point and does the wiggling part again. The return paths alternate between a left and right turn, so the overall path has the appearance of an *8*. As the bee engages in this dance, other workers gather round to witness the dance, which conveys directions as to how to find the food source. The question is, How does the dance provide the needed instructions? Frisch discovered that once a worker witnesses the dance, it immediately leaves the hive and goes directly to the source. So the dance does indeed provide sufficient instructions, and in an ingenious series of experiments Frisch learned the code of the dance.

He first discovered that the direction of the waggle dance indicates the direction of the food, and the length of the wiggling phase indicates the distance. This he showed by systematically moving food sources (dishes of sugar water) around the hive and observing the dance of the first visitor upon its return to the hive. It became apparent that the directional cue indicated the direction of the food *relative to the location of the sun in the sky.* Thus, the sun compass of the honeybee was discovered.

We've already seen how use of a sun compass requires a biological clock, because the sun moves through the sky. Thus, in order to head south, the bee must know not only where the sun is but also what time of day it is. To demonstrate that bees take the time of day into account, bees were trained to find a food source that was located, say, due south of the hive. They were trained on the east coast of the United States. Then they were flown overnight to the west coast, across three time zones. In order to encourage the use of the sun compass, the "training hive" (the one on the east coast) and the test hive (the one on the west coast) were placed in areas devoid of any conspicuous landmarks that the bees might use. Now, if the bees use a time-corrected sun compass, they should show a particular pattern of direction "errors" when they begin their foraging from the test hive. This is because of the time difference between the locations of the two hives. Let's say the bees were permitted to emerge from the test hive at 9 A.M. California time. The sun is in the eastern sky. The bees are still on "east coast time," however: their circadian clock tells them it is noon. During training, the bees learned that at noon, they should fly toward the sun, since at noon the sun is in the southern sky. When they emerge from the test hive, they therefore fly toward the sun. However, since it is really 9 A.M., the sun is in the eastern sky, so the bees fly *east* to look for the food—just as we would predict if they used a sun compass! (See figure 11.4.)

Here's where the really interesting part begins. How do the bees "perceive" the location of the sun? The question is not as simple as it might sound. First, we must remember that our measure of the apparent location of the sun is provided by the direction of the waggle dance. From our perspective (and no doubt from the bees' as well), this is the true importance of the waggle dance, for now we can perform experiments to determine exactly what "the direction of the sun" means to the bee. The honeycombs inside the hive are usually oriented vertically, so the

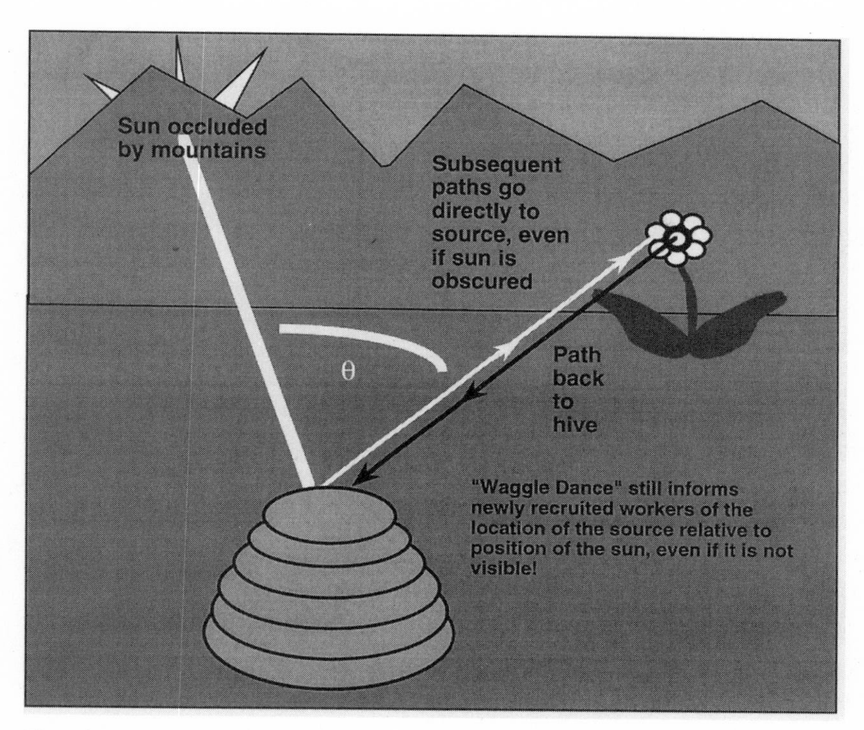

Figure 11.4
Bees specify the direction of a food source relative to the sun, even if the sun is not directly visible.

dance is ordinarily performed on a vertical dance floor. When doing the waggle dance on a vertical dance floor, straight up means "directly toward the sun." If the source is, say, 20° to the right of the current position of the sun, then the bee's dance is oriented 20° off vertical. Now let's suppose that we have placed a little window at the top of the hive—a little honeybee skylight, as it were. Well, we might not be surprised to find the bees can correctly signal the location of the food source when the sun is actually visible through the little window. That in and of itself doesn't get you the Nobel Prize. What Frisch discovered is that the bees "know" the position of the sun even if it is *not* visible through the window, and this important observation led to the discovery of a new compass—one that is based on the location of the sun but that does not require seeing the sun.

It is not hard to imagine the benefits of a sun compass that does not actually rely on direct visibility: the sun isn't always visible (indeed, at certain times of the year in New England, it is almost never visible). On partly cloudy days, there might be patches of blue sky, but the sun itself could be obscured. This, it turns out, is no problem for these insects. They can tell where the sun is, simply by seeing a little blue sky!

How do they do it? Well, first, we know that bees don't use the entire visible spectrum for their compass. In fact, they use a part of the spectrum that we can't even see—the ultraviolet portion. Sunlight is composed of a form of energy called electromagnetic radiation. Electromagnetic radiation is one of the fundamental forms of energy in the universe, and much of modern physics has been devoted to studying its nature. In many ways, electromagnetic energy behaves as if it is made of waves (like sound waves or ocean waves), but in other ways it acts like it is made of particles (those things we call photons). This "split personality" characteristic of electromagnetic energy used to upset physicists, but it upsets them much less now. Nowadays, people take a practical approach and think about light as waves when that is most useful, and as particles when that is most useful. It's what is often called the wave-particle duality of electromagnetic energy.

The sun ejects particles of light (photons) into the universe in all directions. Each of those particles has some energy associated with it, and the great German physicist Max Planck discovered that the energy associated with any particle can have only certain discrete values. Of course, photons also behave like waves, and like any wave, each photon has a wavelength. The wavelength is the distance between peaks of waves. The relationship between the energy of a photon and its wavelength was provided by Albert Einstein in 1905, and follows the relationship $E = hf$, where E is the energy, h stands for a universal constant of physics called Planck's constant, and f is the frequency of the light (1/wavelength). Thus, the shorter the wavelength, the higher the frequency, and the higher the frequency, the greater the energy. Particles of electromagnetic radiation have different energies and different wavelengths. The different wavelengths constitute what we refer to as the electromagnetic spectrum. (See figure 11.5.)

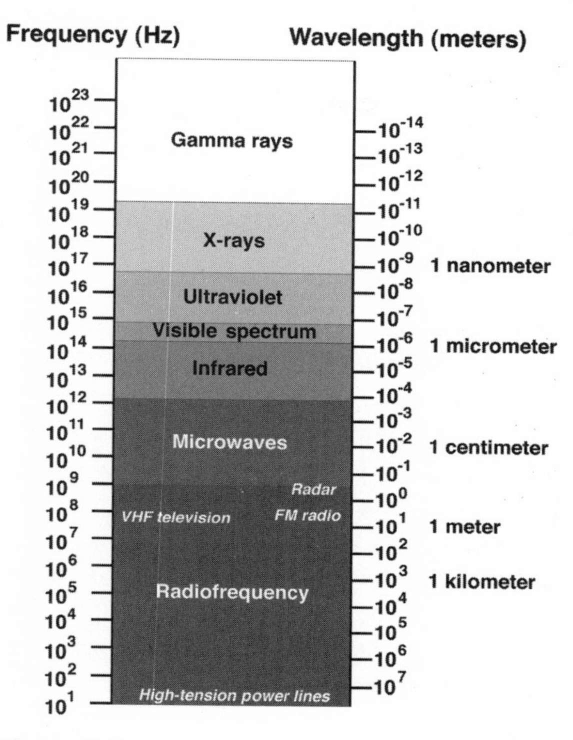

Figure 11.5
The electromagnetic spectrum.

What we call light is really a very small portion of the electromagnetic spectrum—in humans, electromagnetic energy is visible for wavelengths from about 425 nm (nanometers; 1 nanometer = 0.000000001 meter) to 660 nm. This tiny portion of the entire electromagnetic spectrum is what we call the *visible spectrum,* because we can see this light and the different wavelengths appear to have different colors. Light below about 480 nm looks blue, around 510 looks green, 570 looks yellow, and 620 appears red. Even though we are sensitive only to wavelengths of light ranging from about 420 to 660, photons can have wavelengths that range from 0.00000000000001 meter up to 100,000 meters! Obviously, visible light is a *very,* very small part of the electromagnetic spectrum.

The part of the spectrum that has wavelengths just a little shorter than those visible to us is called ultraviolet light (UV; beyond violet). These are the photons that produce sunburn, and in large doses are harmful to our skin and our eyes (thus the emphasis on UV protection in sunglasses and tanning products). We've mentioned photopigments in the context of the optical pumping model of magnetoreception. These are large molecules found in specialized receptor cells. The job of these photoreceptor cells is to absorb photons and convert those absorptions into electrical signals. These electrical signals are the currency of neural information processing, and underlie all of our sensory experiences (and probably our thoughts, memories, and emotions, for that matter).

The pigments do not absorb photons of different wavelengths equally well, however. In the normal human eye, there are four different types of photoreceptive pigment molecules. Three of these work in normal lighting conditions and support color vision. The fourth works only in very dim light. Thus, while we can see in dim light, we can't appreciate different colors in dim light—everything appears to be some shade of gray. For color vision to be possible, biological visual systems must have more than one pigment molecule, and those pigment molecules must be maximally sensitive to different wavelengths of light. The human pigments have peak sensitivities to red, green, and blue light. That is why any color can be produced by the appropriate mixtures of red, green, and blue light—the fundamental principle underlying color television.

For some time, researchers thought that honeybees might be colorblind, because they apparently cannot discriminate red and green. Later work, however, showed that, like us, honeybees have three pigments, but the peak sensitivities of their photopigment molecules are different from those of human photopigments. The three photopigments of the honeybee's eyes have peak sensitivities to green, blue, and ultraviolet. Thus, honeybees obviously see the world differently than we do. Red flowers will not have a color that is distinct from the green leaves, but may look much darker than the leaves. In contrast, dark blue flowers will look brilliant to honeybees. In an interesting example of parallel evolution, flowers appear to have adapted to the UV sensitivity of bees. Almost all flowers reflect some light in the ultraviolet portion of the spectrum, and many use ultraviolet wavelengths to indicate where on the flower the nectar is located. These markings are called nectar markers. Examples of these adaptations are illustrated in figures 11.6 and 11.7.

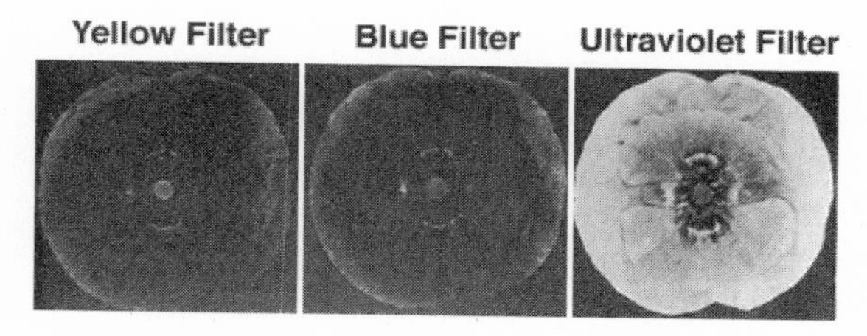

Figure 11.6
Photographs of a poppy blossom taken through filters that pass only yellow light, blue light, or ultraviolet (UV) light. Since the blossom is red, it looks very dark when photographed through a yellow or blue filter. The same blossom reflects a great deal of UV light, however, and therefore looks quite bright when photographed through a filter that passes only UV light. The UV-sensitive photoreceptor cells would respond vigorously to a red poppy blossom.

Figure 11.7
Photographs of a type of violet that has a "nectar marker." In this case the nectar marker is the dark central region of the UV image on the right. This means that the nectar marker reflects little UV light relative to the rest of the petals. This dark "nectar marker" is invisible to us.

What does the honeybees' sensitivity to UV light have to do with their sun compass? We've seen how bees do not need to view the sun directly in order to know where it is—they simply need to view some blue sky. Karl von Frisch asked a simple question concerning this ability: Do the bees use a particular part of the spectrum when they judge the position of the sun? He found an answer by placing different filters over the little skylight window in the beehive. These filters allow light of certain wavelengths to pass, and block out all the others. What he found was that the bees use the UV part of the spectrum. If the UV light rays are present, then they are fine, and their waggle dance is appropriately oriented. If UV light is not present, the bees cannot tell where the sun is, and they reveal this fact in their misoriented dance. So we already see some of this

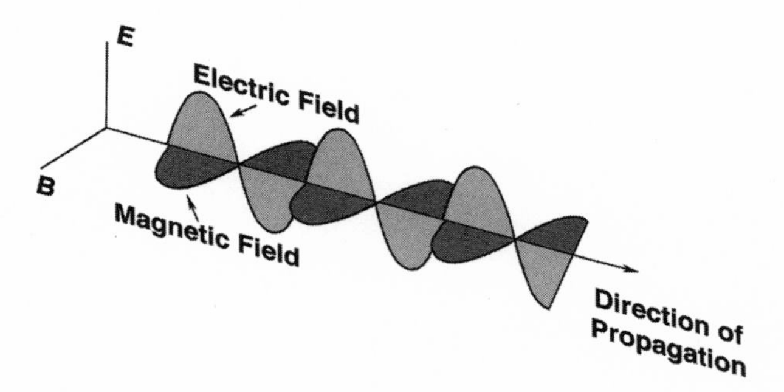

Figure 11.8
Light as a transverse wave. Note that the electric field (E) and the magnetic field (B) are perpendicular to the direction of propagation and perpendicular to one another.

remarkable story: the sun compass of bees depends on UV light—a part of the electromagnetic spectrum that we cannot see.

But there is more, much more. For instance, how does the UV light tell the bee where the sun is? To appreciate the answer to that question, we must delve a little farther into the physical properties of light. Once again, we are interested primarily in the wavelike properties of light.

When we considered sound waves in the context of biosonar systems, we saw that sound is propagated by the motion of molecules: molecules of gas in air, and molecules of water in the seas. Those particles move in the same direction as the wave is moving. Waves of this type are called longitudinal waves. Light waves are a little different. Their motion is perpendicular to the direction of propagation, so light waves are called transverse waves.

If we consider a photon of light traveling from left to right, as indicated in figure 11.8, there are really two waves that are closely related: an electric field vector and a magnetic field vector. As indicated in the drawing, both of these vectors are perpendicular to the direction of propagation, and they are also perpendicular to one another. Since the magnetic vector and the electric field vector are always at right angles with respect to one another, we can consider only one; by convention, the one we focus on is the electric field vector, which is often called the E-vector. The essential point for our purpose is that each photon has one and only one E-vector. We can think of the photon as a tiny Frisbee or

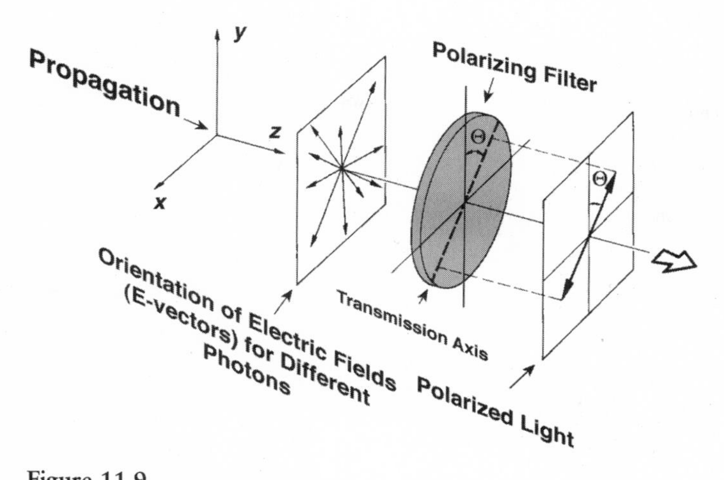

Figure 11.9
Illustration of the polarization of light. Only those photons whose E-vectors match the transmission axis of the filter can pass.

disk that is flying in a straight line toward us—the plane of the Frisbee could be in any orientation. Imagine that each Frisbee can adopt only one orientation: vertical or horizontal or anything in between.

We can think of rays of sunlight as millions of little Frisbees flying toward Earth. The E-vectors are pointing in all possible orientations—some up and down, some sideways, some at oblique angles. Light like this is said to be unpolarized, because the E-vectors are scattered in all orientations. Imagine that the light now has to pass through a grid like a little picket fence. The slots in this imaginary picket fence are oriented vertically. What happens when the unpolarized light strikes this fence? Only those photons whose E-vectors are oriented vertically will be able to pass. The rest will "get caught," and will not pass through. So all the light that is left has the same E-vector orientation, and we say that the light that has passed through this filter is *polarized light* (figure 11.9).

Why are we talking about polarized light? Because Earth's atmosphere takes unpolarized sunlight, and polarizes it. It does this through a process called scattering. Scattering is interesting, and understanding of scattering is fundamental to many natural phenomena. The sky looks blue because of scattering. Sunsets look red because of it. Scattering is thus fundamental to romance, poetry, and, as it turns out, a natural phenomenon that

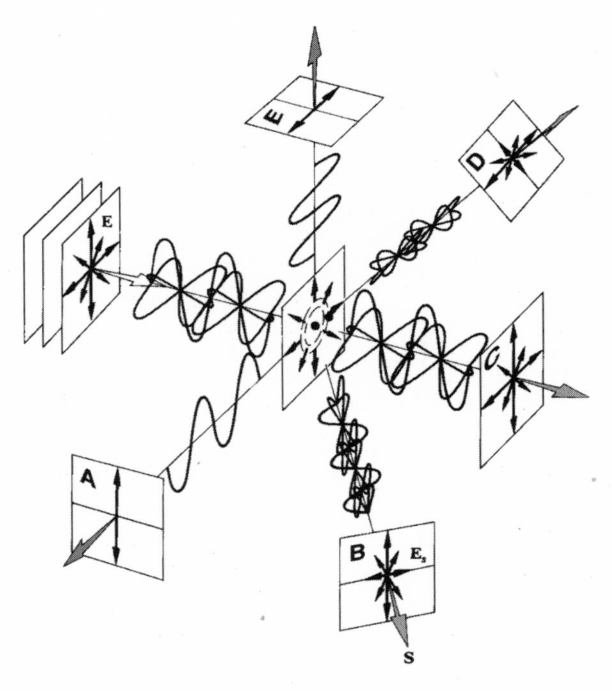

Figure 11.10
Scattering of unpolarized light by a molecule of Earth's atmosphere. The scattered light is strongly polarized in directions perpendicular to the incident sunlight. The scattered light that continues along the original direction remains unpolarized. This means that as you look in different regions of the sky, the light will be polarized by different amounts.

supports the navigational abilities of honeybees. Here is a quick version of how the atmosphere scatters sunlight.

As photons streaming from the sun encounter Earth's atmosphere, some are absorbed by water and other molecules in the air. Because they have just absorbed a photon, these molecules contain an increased amount of energy. They are unstable in this energetic state, and eventually give up the energy by emitting a photon themselves. We call this secondary light "scattered" light. The E-vector of each photon of scattered light is the same as that which produced it. Consider the diagram in figure 11.10.

We refer to the incoming light as the incident light. Notice that the incident light is moving along the z axis, so all the E-vectors are oriented in the x and y planes (the E-vectors must be perpendicular to the direction of propagation). (See figure 11.9.) The scattered light can go off in different directions, but it is polarized. Consider the light that gets scattered in direction A (figure 11.10). Some of the incident light had E-vectors that were horizontal, but that light can't propagate in the direction labeled A, because the E-vector must be perpendicular to the direction of propagation. Thus, the only light that could scatter in direction A is the light that has a vertical E-vector. If we observed the light from vantage point A, the light would be polarized in the vertical orientation. If we observed the scattered light from vantage point B, it would be less polarized than in A, and in C, the light remains unpolarized. Thus the scattered light is polarized to some degree, even though the incident light was not.

The degree of scattering depends on the size of the scattering particles as well as the wavelength of the light. In our atmosphere, blue light is scattered more than red, so the sky away from the sun looks blue. When the sun sets, its light passes obliquely through the atmosphere, so it undergoes a lot of scattering. The reds are spread out the least, so the setting sun looks particularly red. The shorter wavelengths not only are scattered the most, they are also polarized the most. As a result, measurements of the polarization patterns in the sky work best on shorter wavelengths, and ultraviolet light has the shortest wavelengths of all the light that is visible to a bee.

We have seen how the atmosphere polarizes sunlight, but we don't quite yet understand how the pattern of polarized light can tell us the actual location of the sun, even if it is behind a mountain or a building, or is obscured by the canopy of a forest. The key is that the pattern of polarized light in the sky is determined by the position of the sun in the sky. Thus, if you can decode the E-vector pattern in the sky, you can tell where the sun is. (See figure 11.11.)

There is a certain regularity to the celestial pattern of E-vectors: they form concentric rings around the sun. It's also true that this pattern will change as the sun moves through the sky, so we'll have to deal with that. But let's focus on the pattern, given a fixed position of the sun in the sky. Note that if we draw a line perpendicular to the E-vector in any position, that line will eventually intersect the position of the sun. So if we could do this for two different locations, and find where these two imaginary

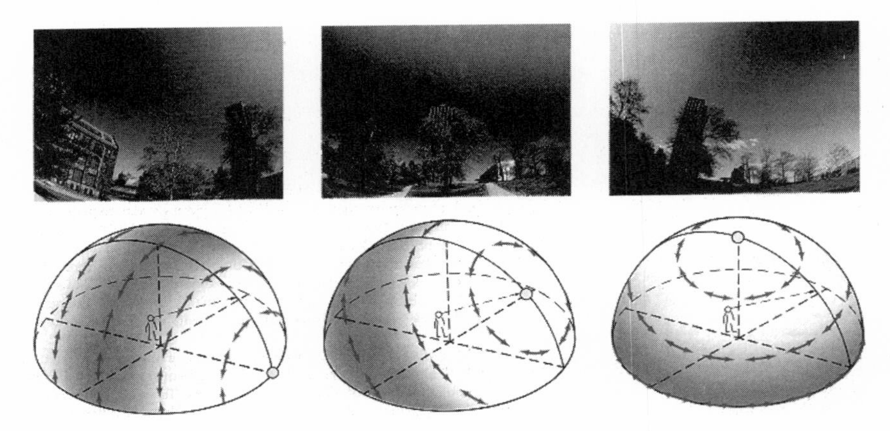

Figure 11.11
Patterns of polarization in skylight at different times of the day. The photographs were taken with a wide-angle lens and a polarizing filter. Without the filter, the brightness of the sky would look uniform to us. But does it look uniform to a bee?

lines cross, we'd know where the sun is! All we need is some analytic geometry and voilà, we're done.

Well, maybe not. Maybe we shouldn't get carried away with the remarkable abilities of these little insects. After all, they have tiny brains. Do we really want to think that they solve the necessary equations? Maybe there is a better way. A simpler way. In fact, a more elegant way. Let's see.

We'll come back to this, but for the moment, let's deal with another mystery in our story. There is a thorny problem that is strongly implied in all of this E-vector business, but that we have deftly avoided until now. How do the bees know what the orientation of the E-vector is? After all, if you look up at the blue sky, you can't see anything but sky blue, maybe mixed with a little smog, depending on where you live. We can't tell the plane of polarization in polarized lights—we can't even tell whether the light is polarized or not. But bees can. How do we know they see the plane of polarized light, and how do we know how they do it?

The waggle dance provides a convenient indication of where in the sky a bee thinks the sun is. While it is also famous as an example of "communication" in the world of insects, it's the message (apparent location of sun) that is most relevant to our present concern. Bees can tell the location of the sun without actually seeing it. What's more, Frisch was able to establish that the sun compass of the honeybee depends on the ultraviolet part of the spectrum. He established these facts by placing filters that either pass or remove UV light within over little windows in the top of the hive—if UV light was present, the waggle dance was accurate; if UV light was removed, the bees did not orient the waggle dance accurately. We also know that the polarization of light scattered by the atmosphere is strongest in the UV part of the spectrum. Frisch also showed that it is the plane of polarization that is important in the honeybee's ability to determine the location of the occluded sun. Placing a depolarizing material on the skylight in the hive disrupts the orientation of the waggle dance, so we know that without plane-polarized UV light, the bees cannot identify the location of the sun simply by viewing blue sky. Thus, we know that the bees can identify the polarization plane of UV light. But how?

As usual, much of the secret lies in the photoreceptor cells themselves. The pigment molecules that begin the visual process by absorbing a photon have a preferred E-vector axis for the incident light; that is, the molecules absorb photons most effectively when the electric field vector is aligned with the molecule. Photons whose E-vectors are not aligned with a given molecule's preference are not absorbed as readily. In humans, the photopigment molecules are sequestered in little disks that lie in the receptor cells. In each disk, the molecules are oriented randomly, so that no matter what the orientation of the E-vector is, there are pigment molecules in each receptor that are "tuned" to that orientation. As a result, human photoreceptors do not respond differentially to plane-polarized light, and light beams that are polarized in different orientations therefore look the same to us.

As shown in figure 11.12, the eyes of insects are very different from our eyes. Our eye is often compared to a camera. We have a cornea and a lens, which acts to project a focused image onto the back of the eye. The back of each human eye contains about 120 million photoreceptor cells. In addition to the photoreceptor cells, there are millions of neurons in our eyes that perform the initial processing of this projected image. The photoreceptors and the associated neural network constitute the

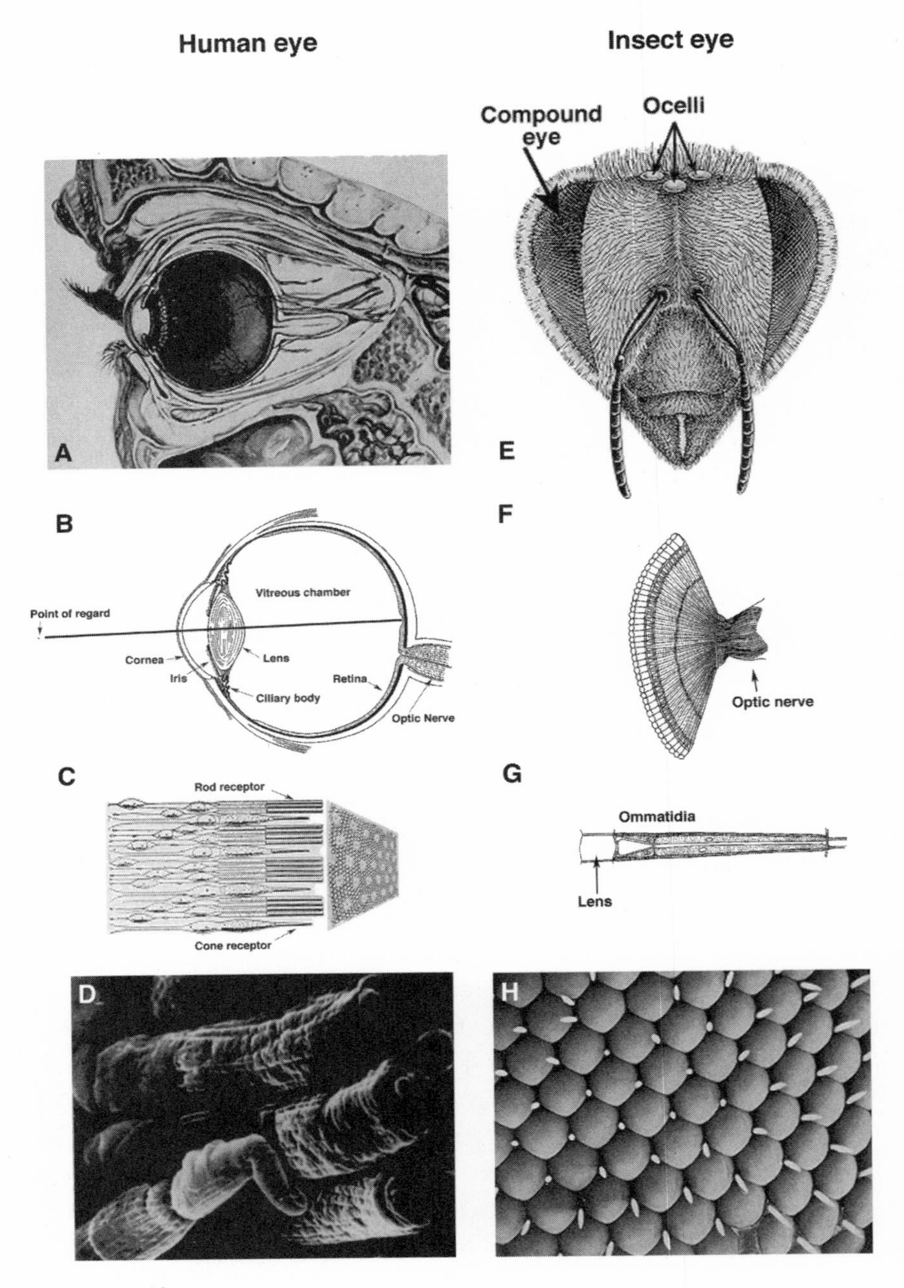

Figure 11.12
Comparison of human and insect eyes. The human eye has one cornea and one lens (shown in B), which serve to project the image onto an array of photoreceptors (a small section is shown in C). A scanning electron micrograph of the tips of rods and cones is shown in D. The cone tip was bent during tissue processing.

retina. All vertebrate eyes have this general arrangement: a cornea and a lens focus an image and project it onto the retina. The retina then sends signals to the brain via the optic nerve. The early processing of the image takes place in the retina, but we should recognize that much of our experience of the visual world depends upon processes that take place within the brain. We might even say that vision begins in the eye but occurs in the brain.

Now let's compare this arrangement with that of a typical insect. Insects have what are called compound eyes. This is a much more "primitive" arrangement. The surface of a compound eye has the appearance of an elaborately cut gemstone. Each "facet" contains an individual lens and a small number of photoreceptors. Each of these facets is called an *ommatidium,* (plural, ommatidia), and each has a specific field of view. Adjacent ommatidia have adjacent fields of view, so together, they cover the entire visual field of the insect. There are many fewer ommatidia in an insect eye than there are photoreceptors in the human eye. As a result, our visual acuity is much better than any insect's acuity. (See figure 11.13.)

Now let's look at a single ommatidium a little more closely. In a honeybee, each ommatidium contains nine photorereptor cells. We've pointed out that the honeybee has three different types of pigment molecules, one most sensitive to green light, one most sensitive to blue light, and one most sensitive to UV light. Each contains three "green" receptors, three "blue" receptors, and three "UV" receptors. I place the colors in quotes to remind us that each of these receptors responds to a range of wavelengths, and these names refer to their most sensitive color.

The anatomical structure of an insect photoreceptor is very different from that of a human. Insect ommatidia are long and thin (like a human receptor), but the pigment molecules are found in cylindrical structures called microvilli. The microvilli run the length of the photoreceptor cell, and project toward the center of the ommatidia. The arrangement is not unlike a toothbrush, where the bristles are aligned as the microvilli are

The insect eye is shown in E through H. This is known as a compound eye, in which each "facet" of the eye contains its own cornea and a set of nine photoreceptors. The structure that forms each facet is called an ommatidia. In H, we see a scanning electromicrograph of a subset of these ommatidia.

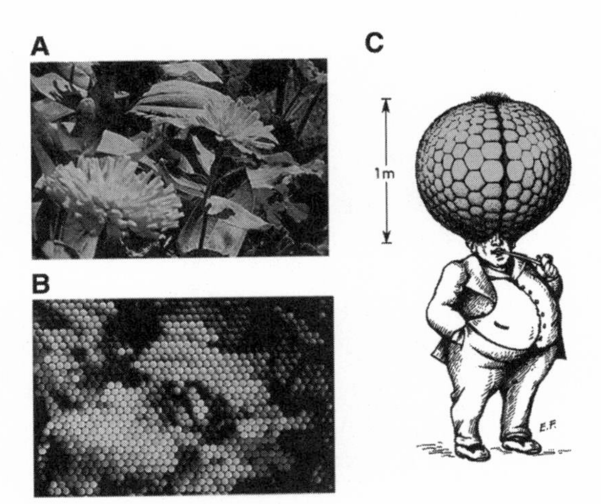

Figure 11.13
Vision with a compound eye. Each ommatidia of the bee's eye "looks" at a specific part of a scene. Since there are many fewer ommatidia in the bee's eye than photoreceptors in the human eye, the "grain" of the visual world of the bee is much coarser than our own. The image in B simulates this coarse sampling by the honeybee's eye. Naturally, the mosaic actually seen by the bee would contain color information, but the appearance of the different colors would be very different than in human sight. In order to have the same acuity as the human eye, a compound eye would require many millions of ommatidia, which would make it impractically large (C).

in the receptor cell. This orderly arrangement of microvilli is the key to the honeybee's sensitivity to polarized light. Unlike the random orientations of pigment molecules in human photoreceptors, the pigment molecules are aligned in parallel arrays in an insect ommatidium. Since the molecules in any one photoreceptor are parallel, each photoreceptor will respond best to a particular plane of polarized light. (See figure 11.14.)

There was a small problem however. It arose when Rüdiger Wehner, a physiologist/zoologist at the University of Zurich in Switzerland, described a curious feature of these ommatidia: they turn like a spiral staircase along their length. This small but important anatomical detail was difficult to demonstrate. After all, the ommatidia are pretty small (only about 0.000180 meter long); it's not easy to tell if something that short runs straight up and down, or if it has a twist in it. But the ommatidia studied by Wehner do indeed turn like a staircase. What is

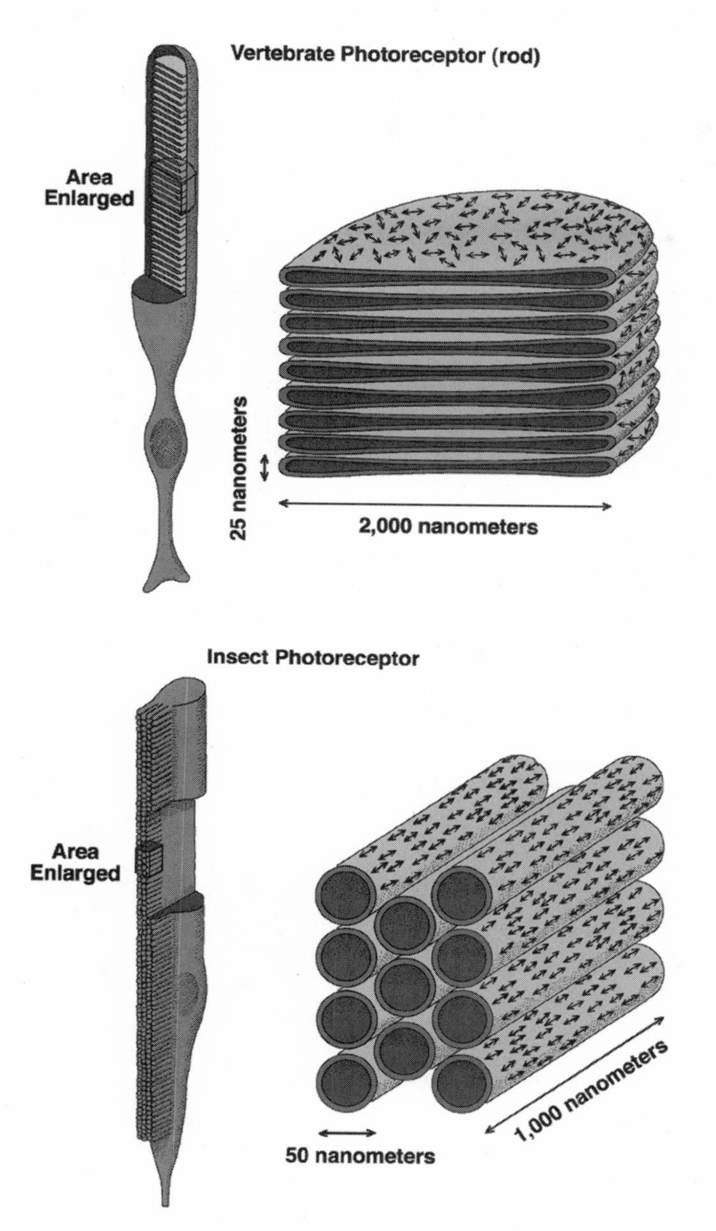

Figure 11.14
Comparison of insect and human photoreceptors. In the human, the orientation of the pigment molecules (represented by the arrows) is haphazard. This is why light polarized in different planes looks the same to us. In the bee, the pigment molecules are aligned within the little filaments called microvilli.

the significance of this? The measurements indicated that the ommatidia turn though an angle of 180° (half of a complete turn), and that posed a problem. The turns in ommatidia *destroy* their differential sensitivity to the plane of polarized light.

In an important part of what is really a wonderful scientific detective story, Wehner and colleagues found a clever way out. They found that in each ommatidium, one of the UV receptors is much shorter than the others (figure 11.15). Since it is shorter, it is twisted through a much smaller angle, and this could preserve a sensitivity to polarized light. Interestingly, half the ommatidia are turned clockwise and half are turned counterclockwise. This renders the two different types of polarization-sensitive UV receptors responsive to different ranges of polarization angles.

So the idea developed that the twists in ommatidia provide the key to the polarization sensitivity of honeybees. According to this idea, the turns were a good thing, because they ensured that the green and blue receptors would not be sensitive to polarized light, and there was no special advantage to be gained by having polarization sensitivity in the green and blue part of the spectrum. The fact that there was only one short UV-sensitive receptor (no. 9) was interpreted as important as well. The reason for this is that the skylight is not perfectly polarized—under natural conditions, only a proportion of the light from one region of the sky is polarized, so skylight polarization is relative rather than absolute. In order to compute the relative degree of polarization, the amount of plane-polarized light must be divided by the total amount of UV light in that part of the sky. Comparing the responses of the long and short UV receptors could accomplish this task.

An apparent order had been established. Challenging problems that had interested scientists for a long time appeared to have been solved. Then the story took another turn. Work continued by Wehner and colleagues (Wehner and Strasser, 1985) found that there were some ommatidia that did not have a short UV receptor and, remarkably, these ommatidia did not twist at all. These straight ommatidia are in the uppermost region of the eye, the part that looks straight up—toward the sky.

Using very small electrodes, the responses of the UV-sensitive receptors of the ommatidia in the dorsal part of the honeybees' eyes were recorded

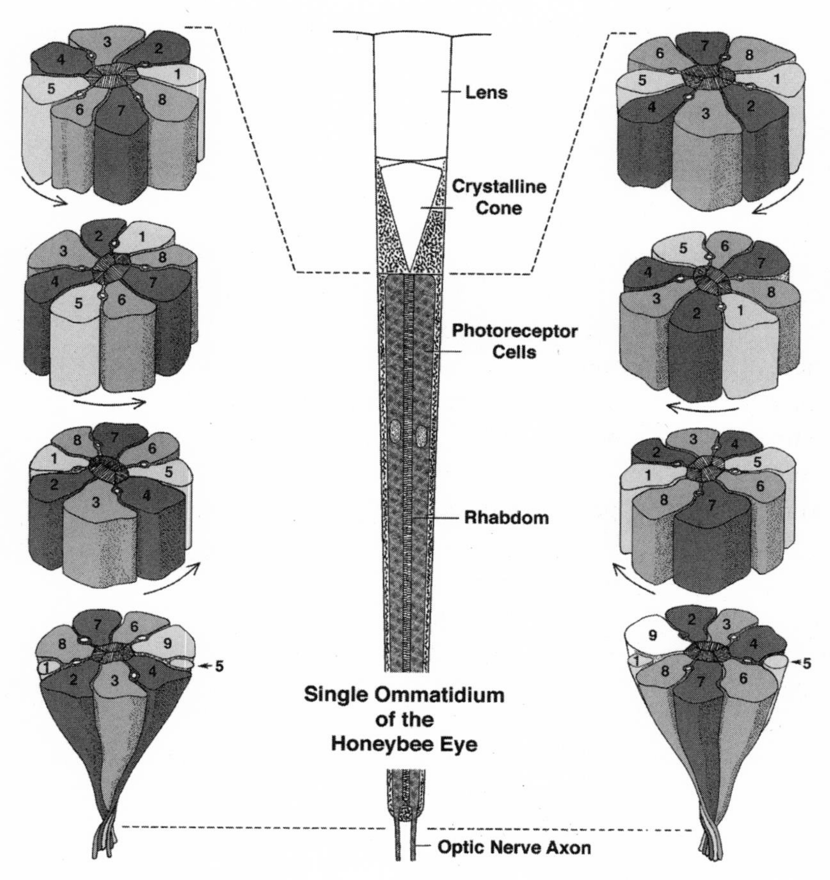

Figure 11.15
Clockwise and counterclockwise twists in honeybee ommatidia. Photoreceptor 9 is shorter than the others (it's seen only in the lowest cross sections at the bottom), and thus could preserve differential sensitivity to polarized UV light.

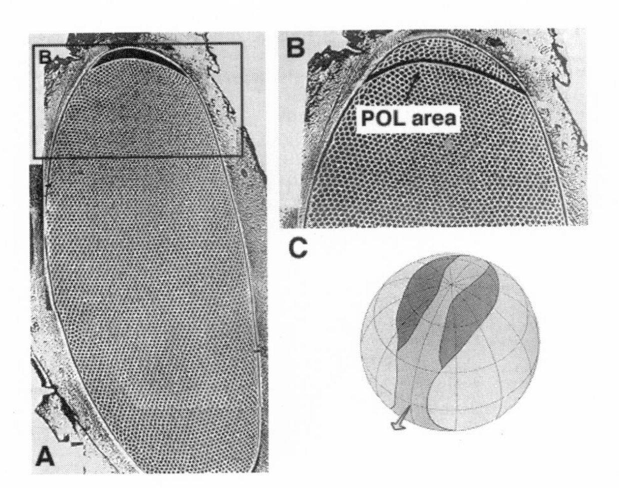

Figure 11.16
The uppermost part of the compound eye of the honeybee has ommatidia that do not twist, are sensitive to UV light, and sample the sky above the bee. There are only about 150 ommatidia in this specialized region of the eye, which Wehner and colleagues call the POL area (POLarization-sensitive area). A is a lacquer replica of the bee's eye. The POL area is the blackened region. B is a closer view of the POL area. C is the estimated field of view of the POL areas of each eye, projected onto a globe whose coordinates represent the sky above the bee. Direction of flight is indicated by the arrow.

by Labhart in 1980. They are extremely sensitive to the plane of polarization of UV light. This specialized region of the honeybee's eye is therefore called the POL area (POLarization sensitive area). There aren't many ommatidia in the POL area, but there are enough to provide coarse coverage of a large region of the sky. These findings suggested that the POL area may be a specialized part of the compound eye that is devoted to analyzing the patterns of skylight polarization. That idea was tested by Wehner and Strasser in 1985. (See figure 11.16.)

The concept behind the Wehner and Strasser's experiment is simple. They simply tested orientation behavior of bees after they had painted different parts of the bees' compound eye. Similar experiments had been performed by Frisch, but they were done before the potential significance of the POL area was recognized. Fortunately, the ommatidia in the POL area are a slightly different color, so once you know what to look for, it is possible to identify the extent of the POL area by microscopic inspec-

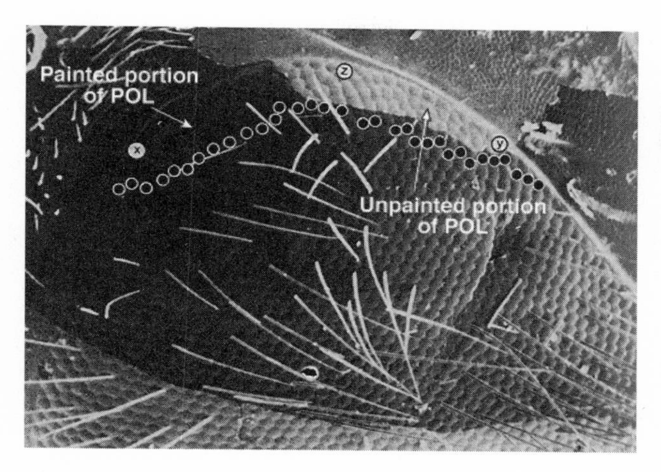

Figure 11.17
Painted and unpainted POL areas. Rüdiger Wehner's experiments show that the POL area is both necessary and sufficient for the bees to use their celestial E-vector compass.

tion. The findings were dramatic and conclusive. When Wehner and Strasser painted the entire POL area of both eyes, the bees were unable to sense the plane of polarized skylight (as revealed by the random orientations of their waggle dance). Orientation was normal if the POL area remained clear but the remaining part of each eye was painted in its entirety. So the POL area is both necessary and sufficient for bees to sense the polarization patterns in the sky. (See figure 11.17.)

How do bees actually use the POL area to orient? Earlier we mentioned that bees can, in principle, identify the location of the sun if they can sample from two different parts of the sky (neither of which need include the sun). We pointed out, however, that in order to do that, some computations would be needed, and it wasn't clear whether we should try to formulate notions as to how the bees peform those mathematical computations.

It now looks as though the key to the bee's E-vector compass is not to be found in neural computations at all. Rather, it appears that the anatomy of the POL area performs most of the work. This is a truly elegant solution to what might otherwise be a very difficult problem. Each ommatidium in the POL area has a preferred polarization angle,

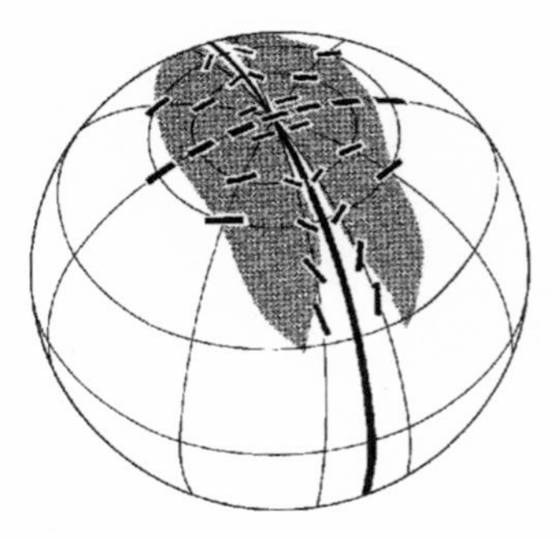

Figure 11.18
Preferred E-vector orientations of ommatidia in the POL area projected onto a map of the sky. These orientations are approximately matched to the E-vectors of skylight in the corresponding part of the sky, so long as the bee is looking in the direction of the sun.

and we know that the pattern of polarization angles in the sky can be used to specify the location of the sun, even when it is occluded. The pattern of preferred polarization angles observed in POL ommatidia is not at all haphazard, but forms a very regular pattern that is the same in all bees. Astonishingly, the pattern is a reasonably close approximation to the patterns of polarization that occur in the sky, especially when the sun is near the horizon (figure 11.18). So the pattern of polarization angles in the skylight might actually be encoded in the anatomy of the bee's eye.

Rossel and Wehner have suggested that as the bee orients in different directions, the match between the pattern of polarization in the sky and the anatomical pattern in the POL area varies. When the match is not very close, each ommatidium "views" a portion of the sky whose plane of polarization does not match that of the receptors, so the response is very weak. When, however, the bee is oriented in the direction of the sun (even though it may not be visible), the polarization pattern of the sky matches that of the receptors, so they all respond vigorously. Thus, the bee only needs to turn until the sky produces the best response in the POL area ommatidia. Then it is facing the sun, and can orient its waggle dance accordingly. This is a remarkable example of "computational anatomy"—what looked like a math problem is solved by an exquisite anatomical adaptation.

III

Electroreception: An Ancient Sense

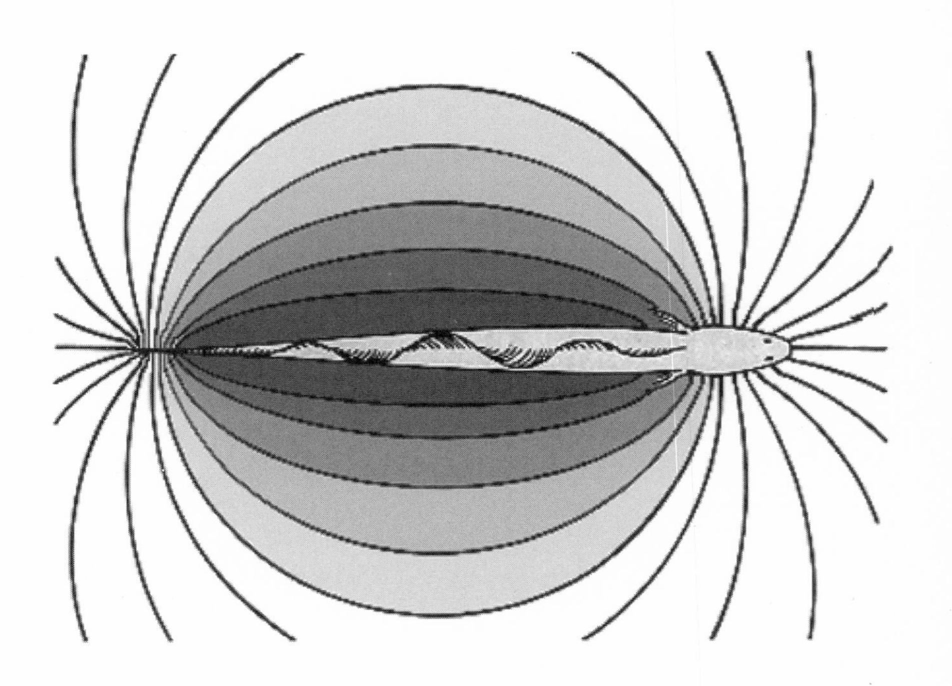

12

The Discovery of Electroreception

Sharks. The primordial terror of the seas. Unchanged for millions of years, sharks long ago evolved into the perfect predator. The ultimate eating machine. We've all seen plenty of images of enormous sharks on the prowl or in the midst of a savage attack. Those piercing eyes that seem incapable of sight. Those razor-sharp teeth. But what you may not know is that sharks have evolved an elaborate sensory apparatus that does not rely on light, or odors, or sound. It is a sensory modality that could not exist in land animals but is found in many marine creatures. It's a sensory modality based on electricity.

Electroreception couldn't be a more efficient prey-detection system, because every living animal that swims in the sea broadcasts its presence by projecting electrical fields into the water. If you're alive and in the water, your body is generating electrical fields. There is no camouflage, no way to hide. And we know that if a form of stimulus energy is available to provide sensory information, life forms will evolve sensory systems that use that information to promote their own survival. In the case of the sharks, electroreception not only helps them detect prey, but may also provide a sense of compass direction.

Since the sharks can sense only electrical fields produced by other animals, they are said to possess a sense of *passive* electroreception. Passive electroreception is distinguished from *active* electroreception, in which a fish produces an electric field and analyzes disturbances in this self-produced field to avoid obstacles, find hiding places, detect prey, and

Figure 12.1
The business end of a Tiger shark.

even find mates. The distinction is analogous to the distinction between sonar and hearing: hearing is passive in the sense that the listener merely receives acoustic energy that is produced by an external event. Sonar is active in the sense that it depends on the echoes of self-produced sounds.

We may note further that there are two general types of electric fish: those that produce strong electric fields and those whose fields are very weak. Although they are essential to the fish's survival, we cannot feel the electric discharge of weakly electric fish. This is certainly not the case with strongly electric species—the discharge of an electric eel can kill a horse! The fact that certain fish—rays and eels—produce strong electric shocks was known to the Egyptians, the Greeks, the Chinese, and undoubtedly to many other ancient cultures.

The ancients did not possess an understanding of electromagnetic energy, however, so they couldn't possibly understand how the production of electric fields was a part of a remarkable sensory modality. Indeed, it was not until the 1950s that the connection between the production of electric fields and their role as a sensory modality was finally established. Even then, though, confusion remained, primarily because we had not yet identified the receptor system that made electroreception possible. Certainly clues were available, but it took some time before their significance was fully appreciated. For instance, a unique type of skin pore had been identified in the torpedo ray by the latter part of the seventeenth century (Lorenzini, 1678), but the functional significance of these receptors remained a source of controversy for 300 years.

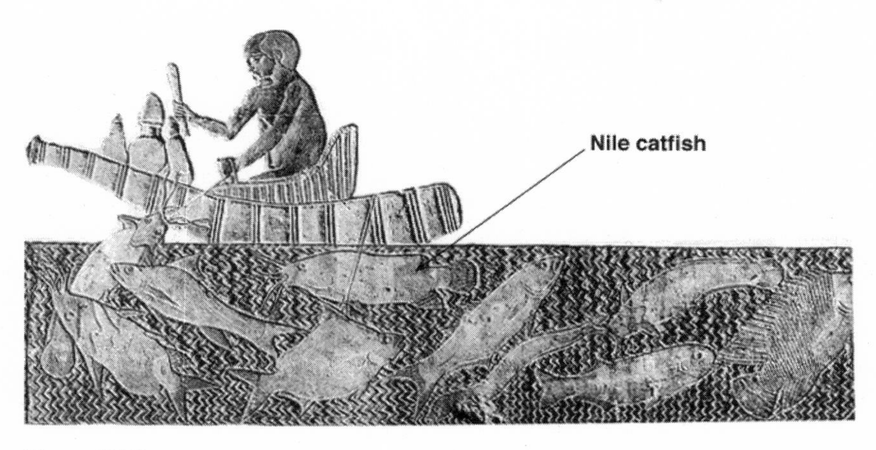

Figure 12.2
The strongly electric Nile catfish is illustrated in this base-relief from the tomb of Ti, circa 2750 B.C.

During the 1950s, many people tried to record the activity of the sensory fibers that innervated these pores (which, in honor of their discoverer, are called ampullae of Lorenzini). Some thought that the ampullae were chemoreceptors, since the sensory nerves that innervate them responded to changes in the ionic or chemical composition of the surrounding water. Others found that the receptors would respond to touch or pressure, and hypothesized that the ampullae were part of a sense of touch, or even a component of a primordial auditory system. Still others thought they were used to sense osmotic changes (changes in the amount of minerals and other material dissolved in the water). So the situation was quite unlike magnetoreception. In magnetoreception, we have demonstrated a sensory modality but have yet to identify the magnetoreceptor unequivocally. In the case of the ampullae of Lorenzini, we had a receptor in search of a modality!

As is so often the case, significant progress had to await certain technical developments. For instance, it wasn't until the late 1700s that we had any real scientific understanding of exactly what electricity *is,* and it wasn't until the 1950s that we developed the electronic instrumentation needed to study the production of electric fields in in living organisms.

Scientific progress in apparently unrelated fields often exhibits a kind of symbiotic relationship, however. A great deal of cross-pollination often occurs. This was certainly the case in the early 19th century, when the physics of electromagnetism and the science we now know as electrophysiology were both in their infancy. As early as 1832, the great British physicist Michael Faraday predicted that ocean currents should produce electric fields as they pass through Earth's magnetic field. Faraday's insight anticipated the idea that the electroreceptive sense of sharks might also serve as a biological compass. And in 1917, two American zoologists, George Parker and Anne van Heusen, noted that blindfolded catfish responded quite vigorously to metal rods placed in the water, but did not respond in a similar fashion to glass rods. Parker and van Heusen recognized that this effect was probably due to the electrical fields generated by chemical reactions between the metal rods and the water; they knew that glass rods could not produce these fields. But at the time, no one foresaw the biological significance of a sensitivity to these galvanic fields, because no one realized that all sea creatures also produce them.

In addition to this scattered evidence that some fish can *sense* electrical fields, it was well known that certain species of fish *produce* electric fields. We've seen that strongly electric species were known to the ancients. The Nile catfish pictured in figure 12.2 can produce shocks of up to 350 volts. A tile mosaic from Pompeii illustrates the Mediterranean torpedo ray, a strongly electric species that is a close relative of sharks (figure 12.3). During the 19th century, the study of the torpedo ray contributed mightily to the scientific understanding of electrical phenomena in living things, and played a central role in the development of modern electrophysiology (Wu, 1984).

For instance, we have already noted that throughout most of the 1700s, it was thought that muscle contractions were produced by the flow of a fluid from the nerves to the muscles. This idea, which originated with René Descartes, seemed plausible enough. After all, when you flex a muscle, it seems to get bigger, so it would be reasonable to suggest that the muscle had been inflated. Yet investigations of the torpedo ray strongly suggested that their electric discharges originated from organs that appeared to consist of a modified form of muscle tissue. Lorenzini recognized that the electric organ of the torpedo ray was probably evolved from muscle. It was also thought (correctly, so far as we know) that fish were the only vertebrates endowed with electric organs, and that

Figure 12.3
A mosaic from the ruins of Pompeii depicts a fight between an octopus and a lobster. A torpedo ray (above octopus) is among the onlookers.

these curious specializations could be found in different body parts in different species. The electric organ of the torpedo is found in its fleshy "wings," whereas that in the electric eel is found in the tail. In both cases, however, the electric organs strongly resemble muscle tissue.

The functional significance of these powerful discharges seemed clear enough: they must surely be an effective defense mechanism, and could easily be used as a weapon to stun potential prey. In this context, the discovery that there were "weakly" electric fish as well as "strongly" electric species appeared especially puzzling. Weakly electric species had been found in the murky rivers of Africa and South America, and the two groups did not appear to be closely related.

The appearance of weakly electric organs in species that were not closely related was a source of concern for Charles Darwin. He saw the whole issue as problematic for his theory of evolution. What was the problem? Well, there were really two problems. One was the apparent uselessness of a weak electric discharge. In contrast to the seemingly obvious evolutionary advantage of a strong discharge, what possible selection pressure could have led to the evolution of an electric organ whose discharge was so weak you couldn't even feel it? After all, a central tenet of Darwin's theory of evolution claimed that the world was a harsh place, and that only the most "fit"—those best adapted to their environment—would be "favored races in the struggle for life." All anatomical forms and specializations were assumed to have a reason behind their existence—a purpose. And the purpose of the "weak" electric organ was something of a mystery. In his landmark work, *The Origin of Species by Natural Selection,* Darwin wrote: "The electric organs of fishes offer another case of special difficulty, for it is impossible to conceive by what steps these wondrous organs have been produced. But this is not surprising, for we do not even know of what use they are" (p. 178).

He went on to ascribe a defensive role to the organs of strongly electric species, but like everyone else who considered the problem, Darwin was stumped by the weakly electric species. Based on the anatomical evidence, Darwin agreed that these organs evolved from muscle. Indeed, he accepted the view that there was an intimate relationship between muscle contractions and the electric discharge, and quotes a Dr. Radcliffe as insisting that "'in the electrical apparatus of the torpedo [referring to the torpedo ray] during rest, there would seem to be a charge in every respect like that which is met with in muscle and nerve during rest, and the discharge of the torpedo . . . , may be only another form of the discharge which attends upon the action of muscle and motor nerve'" (pp. 178–179).

Clearly a great scientist and a masterful observer of nature, Darwin had an intellect that left little room for arrogance. He did not mistake his inability to perceive the purpose of the weakly electric organ for a genuine lack of utility. He clearly stated his belief that science would eventually determine the function of weak electric discharges. Darwin's faith was well placed, although it took another 90 years to establish the role of weakly electric organs in electroreception.

The mysterious and elusive functions of weak electric discharges are not what bothered Darwin the most, however; it was their evolution in apparently unrelated species. He wrote:

> These [weak electric discharge] organs appear at first to offer another and far more serious difficulty; for they occur in about a dozen kinds of fish, of which several are widely remote in their affinities. When the same organ is found in several members of the same class, especially if in members having very different habits of life, we may generally attribute its presence to inheritance from a common ancestor. . . . So that, if the electric organs had been inherited from some one ancient progenitor, we might have expected that all electric fishes would have been specially related to each other; but this is far from the case" (p. 179).

Darwin went on to describe what evolutionary biologists now call convergent evolution: the independent development of the same (or similar) organs in unrelated (or at least very distantly related) species, due to the operation of the same selection pressures. Sometimes evolution provides the same solution to an environmental challenge more than once. Biosonar is one example we have already encountered. Surely bats and dolphins represent species that, in Darwin's words, have "widely remote affinities." Electroreception is another example of a sensory system that appears to have evolved more than once.

A hundred years after Darwin, we finally get a clear view into the secret world of electric fish. H. W. Lissman, a zoologist at Cambridge University, was one of the early pioneers. He wrote a series of articles describing the electrical discharge of several species of weakly electric fish found in the Black Volta River in West Africa. In those articles, Lissman suggests that the fish may use the discharges to navigate and to detect prey. He describes experiments designed to investigate the hypothesis, and the results that clearly indicated these fish use their discharges as part of a sensory system. That system represents an entirely new sensory modality: *active electroreception*. Lissman is sometimes credited with the "discovery" of electroreception. But scientific progress always builds upon—indeed, depends upon—the work of those who went before. In a very real sense, scientific discoveries are always collective achievements. But Lissman's papers do present the first coherent statement of the newly hypothesized sensory modality.

As Darwin had almost 100 years earlier, Lissman pointed out that electric organs are "remarkable adaptations, [but] . . . the problem of

Figure 12.4
A South American species of weakly electric fish.

their evolutionary history . . . *still* awaits a satisfactory answer" (emphasis, mine; Lissman, 1958, p. 156). Lissman agreed with Darwin (and virtually everyone else who had written on the topic) that the strongly electric organs were probably an effective defensive weapon that may also have some value in hunting. Even in 1958, however, the significance of those pesky weak electric discharges remained obscure.

Lissman was clearly skeptical of suggestions that had previously been offered. One was that there was was no environmental selection pressure in the conventional sense, but weakly electric organs evolved in response to "a real inner stimulus working independently of outer conditions" (Dahlgren [1910], as quoted in Lissman, 1958, p. 156). That idea has a certain mystical air about it. An "inner stimulus" that is "independent of outer conditions"—Darwin certainly would have cringed, and Lissman wasn't buying the "inner stimulus" theory either. He also doubted the accuracy of several "outer conditions" that had to that point been considered. These included the ideas that the fish use their weak electric discharge to (1) stun very small prey or (2) achieve some form of protective mimicry by "simulating" strongly electric fish.

So Lissman began his own efforts to understand the purpose of the weak discharge in these unusual fish. And, as figures 12.4 and 12.5 amply demonstrate, they are nothing if not unusual. He noted that the fish are predatory and that they live in very murky water, where vision seems of little use. He described the cannibalistic nature of many of these species, and reasoned that somehow individual fish can detect the presence of

Figure 12.5
The head of a South American electric eel. The pores are probably openings of ampullae of Lorinzini.

another fish at some distance (apparently without relying on vision). He suspected that other curious features of these fish might relate in some way to their still cryptic electric discharges. For instance, Lissman knew that many species of electric fish can swim backward, and that many have evolved an unusal swimming style that permits them to keep their tails very still. The electric organs are often located within the tail.

The fish actually are pretty odd-looking. Many lack dorsal fins, and some don't even have tail fins. They can get pretty big, too—the largest specimens of *Gymnarchus niloticus* run over 5 feet in length!

Lissman described his early observations of several varieties of electric fish in 1958. He had collected the subjects in the Black Volta River and its tributaries during an expedition several years earlier. Lissman eavesdropped on electric fish in the wild by placing wire electrodes in the water, connecting them to an amplifier, and listening to the amplified signals using headphones.

I decided to try this myself. I purchased an aquarium tank for my office, and I bought one fish for it. It's not a pretty fish, but it's certainly interesting. Rather than being brightly colored, this fish is a subdued gray with black spots. It's pretty large for an aquarium fish: about 6 inches. It's obviously a catfish: it has six "whiskers" that radiate from around its large, oval mouth. At the aquarium store, it was simply labeled "electric catfish: $10.99." I brought it to the office and placed it in the tank. Then, in what was not my best-planned move, I thought it should have some brightly colored gravel for the bottom—just to give the tank that homey touch. Rolling up my sleeve, I began spreading the gravel evenly over the bottom. Suddenly . . . zap! It got me! Reflexively I jerked my arm out of the tank, splashing water all over the table. It really hurt! Well, maybe it scared me more than anything, but it was a powerful jolt. It felt like a shock from a wall outlet; it was pain mixed with a sense of vibration. My pet electric fish had shocked me! I suppose I should have known better—the sign did say "Electric catfish." But I wasn't sure until that moment whether I had bought a "strong" or a "weak" species! It was strong, all right. It was one of those Nile catfish. I guess I thought the store wouldn't sell you a fish that could actually hurt you, but in retrospect, that was pretty foolish. I knew they had a lionfish for sale, and they are very poisonous. In any case, I thought that if its shocks could actually hurt me, I shouldn't have any trouble recording its electric discharge—the way Lissman had done in the wild backwaters of the Black Volta River.

So I got a biological amplifier, a couple of wires, an audio amplifier, and a speaker. I laid the two wires in the water after connecting them to the biological amplifier. I connected the output of the amplifier to the audio amplifier and to a laboratory computer. I planned on recording the electric signals of my catfish and storing them on the computer. The audio amplifier was connected to the speaker, so I could hear what was going on.

At first, nothing happened. The fish just swam around. If it was generating any electric organ discharge (EOD), I wasn't recording it. Then I put a number of small guppies in the tank. The woman at the store had sold them to me as "feeder fish." That is when things started to happen. At first, the catfish swam around, apparently searching. It searched along the bottom, but it also frequently moved up to the surface. It appeared to be trying to cover the entire three-dimensional volume of the tank. The fish definitely has eyes, but they are very small and don't appear to see very well. Repeatedly, it would swim right by one of the guppies, even within a centimeter, and do nothing. The catfish acted as though it was completely unaware of how close it had just passed to its dinner. And then, as if by chance, its whiskers came in contact with a guppy. Zap! It sounded like a series of clicks in rapid succession—maybe

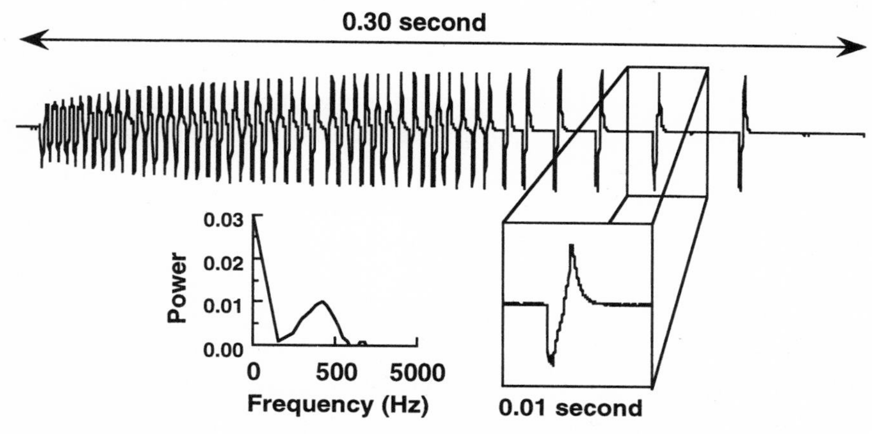

Figure 12.6
Computer-digitized record of the electric discharge of my pet catfish (Sparky) killing a guppy (whom I didn't name). The entire record lasts only 3/10 of a second. Each individual shock lasts about 5/100 of a second. The frequency spectrum of the discharge is illustrated in the lower left panel.

even like machine gun fire. The guppy appeared stunned, and within an instant, it was swallowed whole by the catfish. I caught it all with my computer, which digitized the output of the biological amplifier at a rate of 50,000 samples per second. An example of the records I obtained is shown in figure 12.6.

Catfish stunning guppy. The only time I ever observed an electric discharge from this fish was during predation. Well, I suppose that's not strictly accurate—there was the shock it gave *me*. But I wasn't set up to record it. They apparently use their EOD for defensive purposes and for hunting. I never detected any discharge that appeared to be part of an electric navigational sense—even in the dark, the fish was silent. I sat in the dark with the recording apparatus running for some time.

This absence of electric discharge as the fish simply swims around certainly contrasts with the weakly electric species studied by Lissman. He reported a great deal of electronic chatter in the Black Volta. He could even tell when there was more than one species of electric fish down there.

But electronic surveillance of these fish is one thing; catching them is another. Lissman tells us that he caught most of the fish after they became trapped in pools and bayous that formed when the seasonal floods began to recede. These pools are often teeming with electric fish. You don't need a lot of high-tech equipment then; all you need is a net.

The captured specimens were transported back to Cambridge and were studied over a period of several years. Most of the observations were on *Gymnarchus niloticus* and, when considered together, suggested that the fish are endowed with an electric sense. Here is the crux of Lissman's observations. First, he noted that the fish reacted violently whenever electric current was passed through a wire loop placed in the aquarium water. He also described the curious effects of placing copper wires in the shape of a rectangle on the bottom of the tank. When a fish was placed inside the rectangle, it acted as if it were trapped within an invisible barrier. Every time it approached one of the wires, it would back away. If chased, the fish would eventually "escape" its confinement by swimming over the wire, as close as possible to the surface, while lying on its side. Like Parker and van Heusen, Lissman knew that the wires interacted with the water to produce electric fields. But unlike Parker and van Heusen, Lissman was beginning to think he understood the significance of these galvanic fields. . . .

With respect to the EODs, Lissman reported that they were virtually continuous. The EOD continued after a fish was captured, and also if it was taken out of the water. EODs even continued after death in several specimens.

Several pieces of the puzzle were now falling into place. Being very much impressed with the aggressiveness these animals show toward other individuals of the same species, even in the apparent absence of visual signs, Lissman started to suspect that the fish could detect one another's presence by detecting their self-produced electric fields. He measured the shape and extent of the electric fields that surround the fish: they form an approximation to what is called a dipole source (figure 12.7). Armed with this knowledge and some strong suspicions, Lissman performed an ingenious experiment.

Using wire electrodes inserted into a holding tank, he recorded the discharge of a *Gymnarchus*. He then positioned six pairs of electrodes around the edge of an aquarium tank. The six pairs of electrodes were connected in such a way that the previously recorded "electric message"

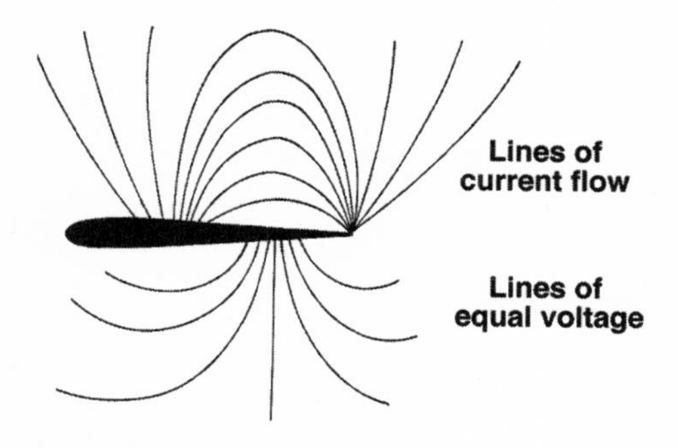

Figure 12.7
The electric field around *Gymnarchus niloticus,* an African species of weakly electric fish.

of a *Gymnarchus* could be played out from any pair. A live specimen was then placed in the tank. The question was, How would the test fish react when the recorded EOD was played through a pair of electrodes? The results were quite clear: the fish attacked whichever electrode was transmitting the recorded signal.

And that's how it started (pretty much). Electroreception. A sensory modality used to find hiding places, avoid obstacles, detect rivals, and even court possible mates . . . all using electricity. Of course, our story is not yet complete. Science is a process; it has no end point. There are always mysteries to be solved. There are, for example, these enigmatic sensory receptors, the ampullae of Lorenzini. Where did they fit into all this? The answer to that question wasn't established until 10 years after Lissman's early work.

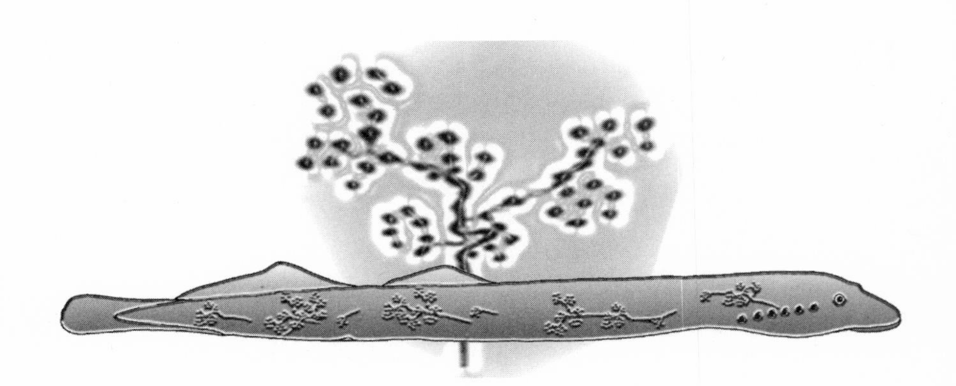

13

The Electoreceptor

Why was it so hard to identify the ampullae of Lorenzini with the sense of electroreception? Well, part of part of the difficulty lay in the fact that ampullae of Lorenzini are found in almost all fish—not just those that produce an electric discharge. Even the primitive lamprey has ampullae of Lorenzini. A close relative of the distasteful-looking hagfish, the lamprey is among the most primordial of all creatures. It has no skeleton, it has no jaw, it has no teeth. It does have a rapacious tongue armed with toothlike appendages on the end that serve as teeth. The lamprey attaches itself to another fish, using its tongue to burrow into the victim's flesh, eating as it goes. One of the earliest branches in the evolution of vertebrates, its only close relative is the hagfish. In any case, the lamprey has ampullae of Lorenzini. Interestingly, hagfish do not.

Perhaps the greatest testimony to the primeval origins of this sensory receptor is that it is found in the coelacanth, a creature so primitive that many thought it was extinct (figure 13.1). For a long time, the only known coelacanths were fossils that were estimated to be up to 400 million years old. Then, in 1931, a live coelacanth was caught in waters close to Madagascar. What an astonishing find! Catching a live coelacanth was considered about as likely as finding a live dinosaur.

Since that remarkable discovery in 1931, about 200 coelacanths have been caught, and it appears that their habitat is confined to the waters off the northern coast of Madagascar, near a small group of islands called the Cormoro Islands. "Coelacanth watchers" do not know how many

Figure 13.1
Perhaps the most primitive vertebrate known, the coelacanth is endowed with ampullae of Lorenzini, and is likely to be an electroreceptive species. Side view is shown in A. In B, many of the ampullae are distributed around this enormous mouth.

of the creatures remain, but there is concern that the species, which apparently has been perilously close to extinction for the last 100 million years, may now be on the verge of finally losing its fight for survival. By any standard of human aesthetics, the coelacanth is not an attractive beast. It has an enormous mouth filled with fiercely pointed teeth, very fleshy fins (it's sometimes called the fish with legs), and a somewhat corpulent body. But the ancient coelacanth has what appear to be ampullae of Lorenzini.

Sharks have them as well, and so do their close relatives the rays and skates (including of course, our now-familiar torpedo ray). The sharks

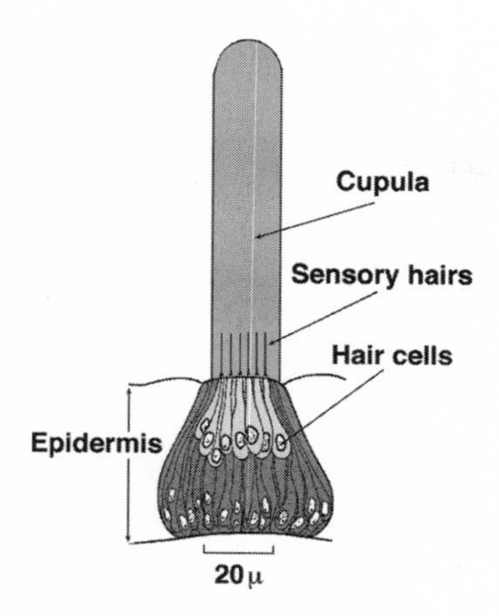

Figure 13.2
Anatomical structure of one type of sensory organ found in the lateral line system of fish. This one responds to water currents that touch the fish's skin (epidermis). Size calibration is specified in millionths of a meter.

and rays are members of elasmobranchs. They are distinguished from "bony fish" in part by the fact that their skeletons are made of cartilage rather than bone. But almost all of the bony fish have ampullae of Lorenzini, too. It was the ubiquity of these receptors that made their function so elusive. After all, if the ampullae are found in elasmobranchs and almost all bony fish, why would anyone think they had anything to do with an electric sense that was apparently specific to a restricted number of highly unusual African and South American freshwater species?

Scientists did not yet appreciate certain key elements to complete the story. But that was soon to change, aided in large part by work that was begun during the early 1960s by a Dutch scientist named Sven Dijkgraaf, (the same Sven Dijkgraaf who had worked with bats 20 years earlier) and one of his students, Adrianus J. Kalmijn. Dijkgraaf did not start out with a specific interest in electroreception. He was, however, interested in an array of sensory receptors that is known as the lateral line organ. The lateralline organ is found on all aquatic vertebrates, and is really composed of a heterogeneous collection of receptor types.

One class of receptor cells has little hairs that protrude into the water. In some hair cell receptors, the hairs are embedded in the mucuslike coating that makes most fish feel slimy to the touch (figure 13.2). In

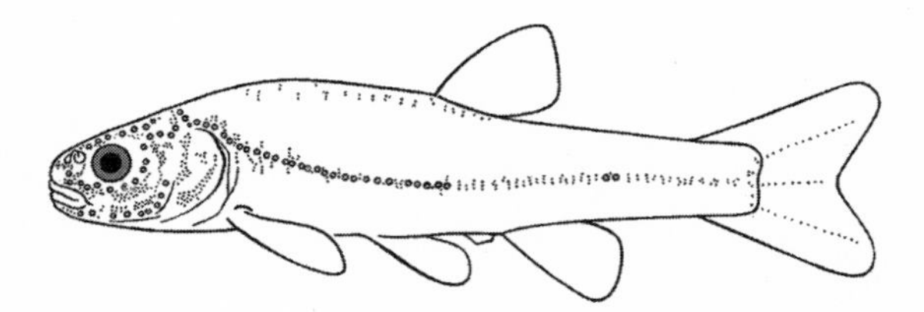

Figure 13.3
Distribution of sensory organs in the lateral line system of a minnow.

others, the hairs extend directly into the water. In addition to these hair cells, however, the lateral line contains ampullae of Lorenzini. These varied receptor cell types are widely distributed over the fish's body. They are found around the head, and extend in a line that runs the length of the animal's body (thus the name; see figure 13.3). When you look at most fish from the side, it is readily apparent that the top and bottom halves have distinctively different coloration. The change in coloring occurs right along the midpoint of the flank—where the receptors are.

The lateral line system provides the fish with what amounts to several different sensory modalities. The hair cells are sensitive to mechanical forces. These forces are produced by water currents and disturbances in the water initiated by movements of nearby objects. They bend the hairs, which in turn causes ionic currents to flow within the receptors. As in all sensory receptors, these current flows are converted into nerve impulses that are conveyed to the brain via sensory nerves. In some ways, these mechanical disturbances have the properties of underwater sound waves, although the receptors respond best to fairly low-frequency vibrations (less than 10 cycles per second, which is lower than normal human auditory sensitivity). In their physical details, these vibrations are not actually underwater sound waves, which are pressure waves that travel at a velocity of 1450 m/sec in freshwater. The stimulus that activates the hair cells is a little different. Rather than responding best to pressure waves, the lateral line organ responds to movements of the water that occur in response to movement of a nearby object. Thus, whereas the hair cells of the lateral line system are often compared with auditory receptors, the stimulus that excites them best is not truly an auditory pressure wave. In some ways, the hair cells of the lateral line organ are

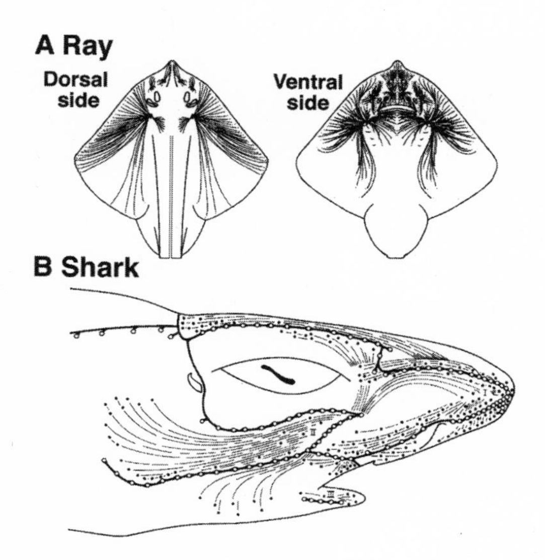

Figure 13.4
Distribution of ampullae of Lorenzini in rays and sharks.

similar to a sense of touch. An apt analogy might be the feeling of a breeze moving across your face. You can hear the breeze, but the feel of it on your skin is a distinctively different modality. In any case, swimming motions of nearby fish can be sensed with this system.

The ampullary receptors are different. First of all, they look different—and as any good anatomist will tell you, form follows function—an adage that unfortunately is often easiest to see retrospectively. Rather than consisting of little hairs that protrude into the water, ampullae consist of receptor cells that are located at the base of a canal. The length of these ampullary canals can be as short as a millimeter, or can be quite long. Figure 13.4 schematically illustrates the location, orientation, and length of the canals found in sharks and rays. It is quite clear that sharks are well endowed with ampullae of Lorenzini, and while sharks have been accused of many things, they have never been accused of producing an electric discharge.

One might think that recordings of the bioelectrical responses of the sensory fibers that innervate the ampullae would establish the class of environmental energy they are designed to detect. After all, visual

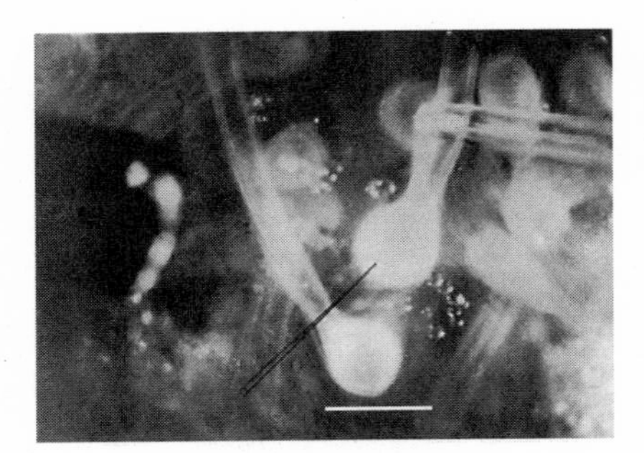

Figure 13.5
Photograph of ampullae of Lorenzini in a living specimen, taken through a microscope. In order to study the physiological properties of the organ, a microelectrode is being inserted into a sensory cell. The figure provides some appreciation for the technical skill this work requires.

receptors respond to light, touch receptors respond to touch. One might have thought it a simple matter to stimulate these receptors with different forms of energy and see which form produces a biological response. The problem was that when experiments of this sort were performed, the receptors were found to respond to a variety of stimuli. Touching them produces responses, and since they do indeed populate the skin, one might reasonably suggest that the ampullae function as a sort of touch receptor. Ampullae also respond to changes in water temperature (thermoreceptors?) and in the salinity of the water (chemoreceptors?). The variety of stimuli that can produce a response in the nerves that innervate the ampullary organs naturally caused some confusion. Which type of stimulus was the receptor actually designed to detect?

The first report that the ampullae were sensitive to electric fields appeared in 1962. Around the same time, Dijkgraaf and Kalmijn were studying the effects of galvanic fields in sharks. In experiments similar to those performed using catfish in 1917, it was found that sharks have a reflexive eye-blink response to very weak electric fields. The fields were about the same strength as the electric fields that Dijkgraaf had earlier found to be a natural consequence of the rhythmic gill movements that occur during breathing in fish. Dijkgraaf was clearly homing in on the solution to the puzzle. He reasoned that since all living fish produce these electric fields, predators like the shark may have evolved a mechanism that can detect them. The eye-blink reflex response to weak electric fields supported this conjecture—sharks always shut their eyes immediately before an attack. It's presumably a reflex designed to protect the eyes. Armed with the knowledge that the ampullae could respond to electric fields of the same strength as those that fish naturally produce, Dijkgraaf and Kalmijn tested the electric field sensitivity of sharks after the ampullary nerves had been surgically severed. They discovered that this operation abolished the eye-blink response to galvanic fields, and thus established an electroreceptive function of the ampullary organs.

In order to firmly establish the importance of electroreception in the daily lives of the shark, however, it was essential to demonstrate the way a shark normally reacts to these weak bioelectric fields. Kalmijn established a behavioral role of electric fields in a series of ingenious studies. First, a live flounder was placed in a shark's pool. The flounder quickly buried itself in the sandy bottom. Like chameleons, flounder are able to change their coloration to match that of their surroundings. All this camouflage was of little avail, however, for the shark, aroused by a delectable fish puree that had also been added to the water, began to search for food. Initially, the searching pattern of the shark was random, but when the shark passed within 15 centimeters of the "hidden" flounder, the attack was quick, accurate, and very effective: the flounder was immediately eaten.

Kalmijn next placed a flounder in an agar container. Agar is a gelatinous substance extracted from a type of algae. It is often used as a substrate in tissue cultures. Electrical currents can pass through it, but it is not transparent to light. The flounder was concealed in this chamber, through which water was pumped in order to keep the fish alive. This concealment was completely ineffective. The shark again searched, and when by chance it came close to the agar, the unsuspecting flounder was immediately attacked and devoured. Placing pieces of dead flounder in an identical chamber did not produce an attack directed toward the chamber. Instead, the shark attacked the end of the tube in which the ventilated water exited from the chamber; this evacuation tube was some distance from the chamber. This last observation proved that scent could provide a cue that would support a shark attack, since the water in this area had a high concentration of fish odor. If the agar experiment was repeated, again using a live flounder but a container wrapped in an electrically insulating plastic film, the shark ignored the chamber.

The final experiment in the series confirmed that electric fields were sufficient to provoke a shark attack. Following a procedure similar to that Lissman had used with *Gymnarchus* ten years earlier, Kalmijn placed electrodes in the sandy bottom of the shark tank. The shark viciously attacked the metal electrodes whenever currents that mimicked the electric fields of a flounder were emitted.

Kalmijn's behavioral experiments firmly established that sharks can detect prey on the basis of the weak electric fields living fish produce in seawater. Thus, two modes of electroreceptive senses have been established: an active mode and a passive mode. In the passive mode, fish can sense the electric fields produced by others. In the active mode, the fish detect disturbances in an electric field that surrounds their own body, a field that is created by their own EOD. Sharks employ only the passive mode. The weakly electric fish use both active and passive electroreception.

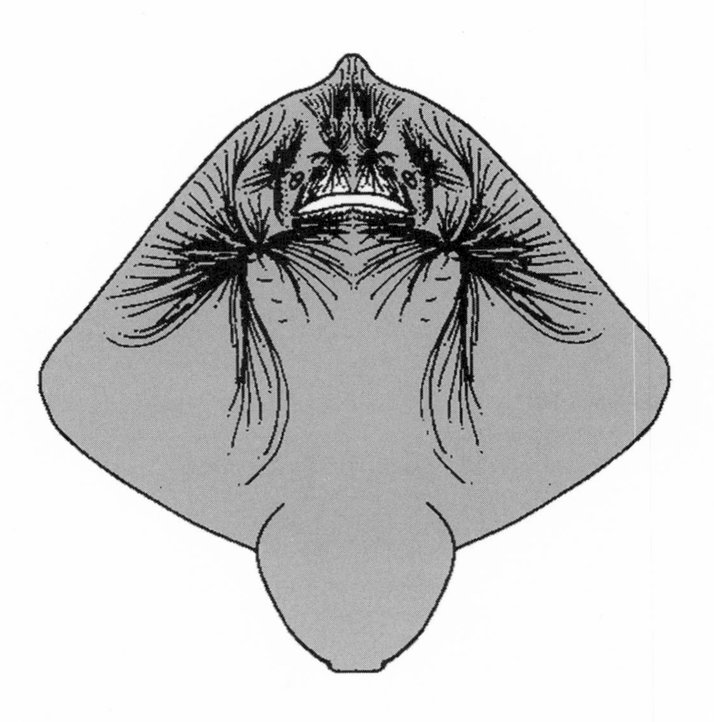

14

The Nature of Electroreceptors

Fish that use active electroreception are endowed with at least two different types of electrosensitive organs. The first is the now familiar ampullary organ. As in sharks and rays, the ampullary receptors are used in passive electroreception. Active electroreception requires a different class of receptor organs, which are called tuberous receptors (figure 14.1). Like the ampullary organ, the receptor cells of the tuberous organs lie at the base of a canal. To the casual observer, they might appear quite similar to the ampullae of Lorenzini. Recordings of the responses of these organs have revealed some important differences, however.

Like all receptor cells, the electrosensitive organs respond only to selected aspects of the electric fields. We say the receptors are "tuned" to a specific dimension of the field, and that ampullary and tuberous receptors are tuned to different properties, or features, of the electric environment. Like auditory receptor cells, the most important dimension on which this tuning depends is the temporal frequency of the stimulus. In hearing it is the frequency of the variations in air pressure that is relevant. In electroreceptors, it is the rate at which an electric voltage changes with time. A helpful analogy can be drawn between the voltages produced by a battery and those produced by an electrical outlet. The voltage produced by a battery is constant over time. A constant voltage produces a constant current, and such electrical sources are called DC (direct current) sources. In contrast, the voltage available from an electrical outlet varies over time. If we were to plot the variations in voltage,

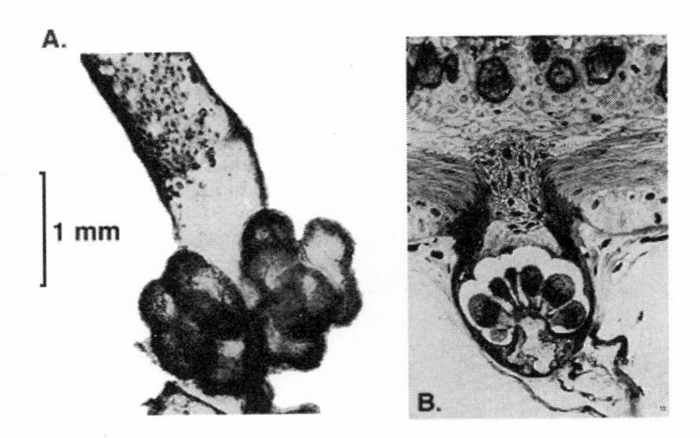

Figure 14.1
A. An ampullary receptor, with sensory cells clustered around the end of a canal. B. A tuberous receptor, with sensory cells arranged in a rosette-like pattern. The canal in tuberous receptors is filled with a cellular matrix that helps determine the frequency selectivity of these receptors.

we would see that one "pin" of the outlet is positive at one moment, and swings negative at the next. The second "pin" varies in the opposite fashion. The voltages at either pin change in a smooth and regular fashion. In fact, the variations are a sinusoidal function of time. Sources like this are called AC (alternating current) sources.

Corresponding to the different types of electric sources, different types of electronic circuits are needed to measure DC and AC electrical fields. For example, one important characteristic of an electric amplifier is the range of frequencies that it amplifies. The devices known as graphic equalizers allow the listener to amplify different bands of frequencies by different amounts. So if you like to emphasize the bass, increase the amplification (often called the gain) of lower frequencies. If you like to emphasize the high notes of a soprano, then increase the gain of the high frequencies. In this way, you can customize your stereo to suit your tastes. In general, the larger the range of frequencies that receive uniform amplification, the more expensive the audio equipment. This range of frequencies is called the bandpass of the system, so in general we can say that the larger the frequency bandpass, the greater the range of sounds a stereo system can play with a high degree of fidelity.

But what if it was important to have a system that transmitted only certain signals and not others? Such a system would be very useful if only a special type of signal was important, and all others were irrelevant "noise." If the important signal was contained within a narrow range of frequencies, then a system whose frequency characteristics matched those of the signal would selectively respond to important information, and the irrelevant "noise" would automatically be eliminated. There is another important benefit when multiple signals must all be processed. Once again, if the signals differ in their frequency components, each can be independently detected by using different detectors, each having a bandwidth that corresponds to one of the possible signals. Thus, the presence of a signal is indicated by the response of one of the detectors, and the identity of the signal is indicated by *which* detector is responding. This is the reason why fish that have active electroreception are endowed at least with two different types of electroreceptor.

The ampullae of Lorenzini are tuned to very low frequencies—from DC to about 10 hertz (1 hertz [Hz] = 1 cycle in 1 second, so 10 Hz = 10 cycles/sec). In contrast, the tuberous receptors are tuned to much higher frequencies—up to several hundred cycles/sec.

Because they are tuned to different frequencies, these two classes of electroreceptors respond to different types of electric fields—they sense completely different aspects of the aquatic environment. The low-frequency sensitivity of the ampullary receptors means they respond best to the electric fields produced by the respiration and movements of nearby fish. The tuberous organs, because they respond only to higher-frequency fields, do not respond to the fields produced by the respiration and movements of other fish. Rather, the tuberous organs of active electroreceptive fishes are specifically designed to sense the fields generated by the electric organ discharge.

How do these fields provide fish with important information about their surroundings? Lissman's early measurements of the spatial distribution of voltages surrounding an electric fish indicated that the electric fields were similar to the pattern of the magnetic field produced by a bar magnet. If you sprinkle iron filings on a piece of paper, and then place a bar magnet under the paper, the iron filings become aligned with the lines of magnetic force in the magnet, and in so doing reveal the shape of the magnetic field. The shape of the magnetic field of a bar magnet is a dipole. Earth's magnetic field is a dipole. The electric fields produced

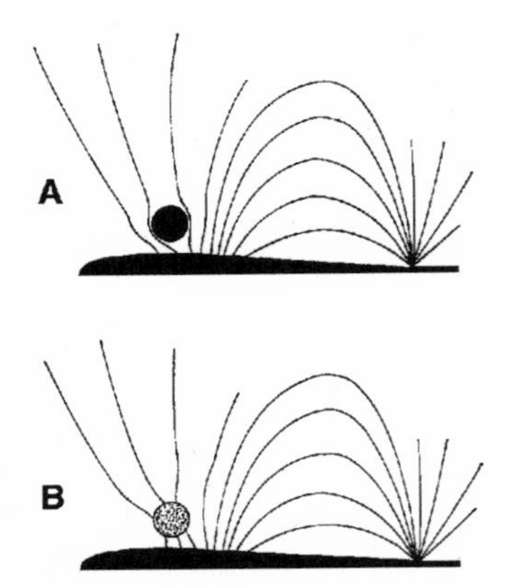

Figure 14.2
Objects near an electric fish distort the distribution of voltages along the fishes body that occur during the electric organ discharge. In A, an object whose electrical resistance is greater than the surrounding water produces a localized reduction in the strength of the self-produced electric field. In B, an object whose resistance is less than the surrounding water produces a localized increase in field intensity.

by the electric discharge organ also are well approximated by a dipole. The shape of the field in open water was illustrated in figure 12.7. In figures 14.2A and 14.2B, we see that nearby objects distort the shape of the electric dipole. These distortions provide the animals with a perception of the nature and locations of objects within the field. The distortions result from differences in the electrical resistance of an object relative to the water. Other fish have a lower electrical resistance than the water, so they will concentrate the lines of electrical force. Many inanimate objects (rocks, etc.) have a higher resistance than the water, so they will produce something analogous to an "electrical shadow." Simplified examples of these effects are illustrated in figure 14.2.

Figure 14.2 also helps demonstrate why the tuberous receptors must be widely distributed over the fish's body—it is a simple consequence of the fact that the distortions can be detected only by simultaneously comparing the strength of the electric field at a variety of different locations. The widespread distribution of tuberous receptors enables electroreceptive fish to detect and localize these electrical "hot spots" and "shadows." The strength of the signal depends upon the distance of the object, just as the sharpness of a shadow is determined by the distance between an object and the surface on which the shadow is cast. In this way, the distortions in the field inform the fish about the approximate distance and location of both animate and inanimate objects.

The entire system appears to rely on a difference between the undisturbed dipole and one that is altered by surrounding objects. This implies that the fish must have a memory, or a template of what the distribution of voltages across its body ordinarily is and it compares that "expected" pattern with the pattern the fish is currently experiencing. As in all perceptual systems, some form of memory is an essential feature.

So active electroreception is based on detecting differences in a self-produced "electric halo"—differences in the spatial distribution of voltages over time. Since the system is based on detecting change, the fish needs to know that the changes detected by the tuberous organs were produced by surrounding objects, not by random variations in the electric organ discharge. For this reason, a premium is placed on regularity and the uniformity of the EOD.

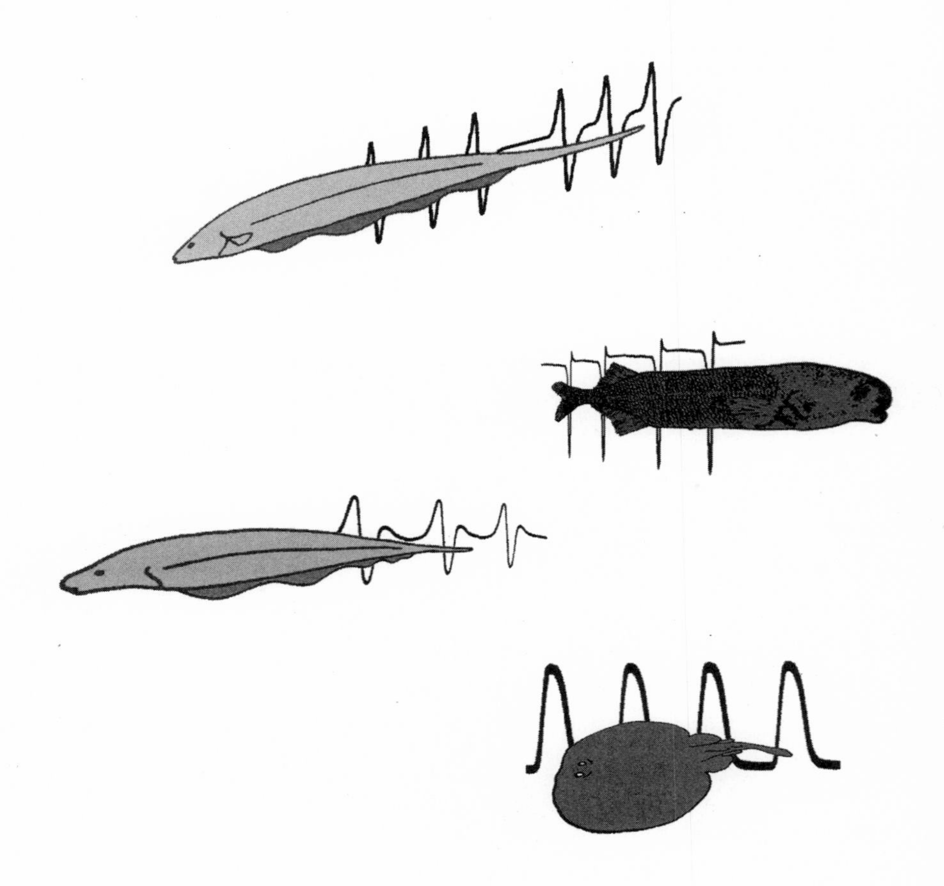

15

The Electric Organ

An analogy can be drawn between biosonar and hearing, on the one hand, and active versus passive electroreception on the other. Sonar is to hearing as active electroreception is to passive electroreception in the sense that both sonar and active electroreception rely on distortions of self-produced signals, whereas hearing and passive electroreception do not. We have noted that the tuberous receptors are specially adapted to respond to the animal's own electric discharge, and this naturally leads us to consider the nature of the electric organ itself.

As pointed out by Dye and Meyer (1986), an engineering solution to the problem of identifying objects by the distortions they produce in an electric field can take one of two possible tacks. One one would be to produce a steady sinusoidal signal (an AC signal at a particular frequency), and analyze the resulting changes in the amplitude and phase of the field produced by an object. The other would be to use a series of very brief, very strong pulses (known as an impulse in engineering), and analyze the distortions produced by the object. Remarkably, both approaches have been adopted in the African and the South American varieties of weakly electric fish. Those species that adopted the former strategy are known as "wave species," and those that adopted the latter are known as "pulse species." (See figure 15.1.) The electric organ discharge of wave species looks very much like a sine wave, although it is not a pure sine wave, and that fact is probably beneficial to the operation of the sensory system. As the name implies, the pulse species produce

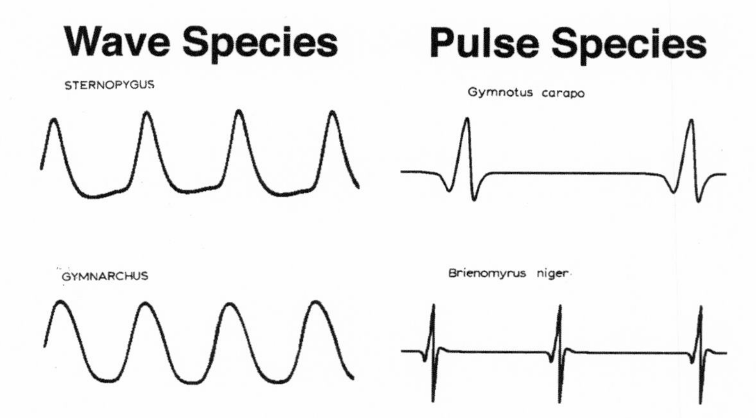

Figure 15.1
Comparisons of the electric organ discharge (EOD) of two wave species and two pulse species.

very short pulses, and the intervals between them are long relative to the duration of the pulse. Pulse species vary the repetition rate of their EOD under different circumstances, and this fact has deep significance to the use of electroreception as a sensory modality and as a tool for communication between individuals. Wave species produce what is essentially a continous waveform—essentially a constant electric "humming."

A typical electric organ is composed of modified muscle tissue, an observation that was well appreciated by Charles Darwin, among others. It is interesting to note, however, that in a few species, the electric organ is not modified muscle tissue but modified nerve cells. Indeed, variation among different species is the rule rather than the exception. Some species possess more than one electric organ; electric eels have three, for instance. One thing does remain constant, however. The discharge of the electric organ(s) of all species is controlled by neurons within the fish's central nervous system, and the operation of these neural circuits has been the subject of intense and creative scientific scrutiny.

There are a number of interesting questions one might ask about the EOD of electric fish. For instance, how do the fish produce EODs of up to 500 volts, using cells that individually produce only very tiny voltages? Part of the answer lies in the fact that if the cells have just the right alignment with respect to each other, their tiny voltages will add. Then, if the cells all discharge at exactly the same moment, their tiny individual voltages can add together. With a sufficient number of cells contributing to the sum, you can produce an arbitrarily large voltage.

For maximum efficiency, the cells in the electric organ must discharge at exactly the same time. Any asynchrony in a pulse species would lengthen the pulse and lower the amplitude; the resulting variability in the EOD would degrade performance of the electroreceptive system as a whole. Asychronies in wave species will also degrade the signal, and reduce overall performance as a result. For both comedians and electric fish, timing is everything.

Consider, for instance, the EOD of one South American wave species. The fundamental frequency of the EOD is 850 Hz. If the fish produced a perfectly regular waveform, it would produce one wave every 0.0001764 second, but the variance around the 850 Hz frequency is only 0.102 Hz. This translates into a time error between successive waves of less than 0.00000014 second! This fish would make approximately 26,805,600,000 pulses each year. If the interval between each pulse was slowed by only 0.00000014 second, and you used the EOD as a clock, your "electric fish clock" would lose only about an hour each year. That is pretty incredible timekeeping for a lowly fish. Now the regularity in these discharges is obviously not to keep time in the everyday sense, but to sense small delays in the electric field produced by surrounding objects. The better the animal's sensitivity to these delays, the more acute the sense. These fish can detect a delay as short as 0.00000040 second, so the regularity of the waveform must be greater than that, or they couldn't tell whether the delay was produced by an object or was in the EOD itself. That is why such a premium is placed on regularity of the EOD in wave species.

The situation in pulse species is somewhat different. These animals can, and do, vary the rate at which the electric pulses are emitted. Part of the reason for this is that sensory information is only obtained in discrete "packets" during each pulse. The situation is akin to vision under stroboscopic illumination: you can see only when the strobe light is on. When confronted with some novel object, pulse species immediately increase the repetition rate of their EOD. This novelty response is reminiscent of the feeding buzz of CF-FM bats. Changes in the repetition rate of the EOD in pulse species also serve to communicate messages between individuals of the same species. We'll consider these communicative processes shortly. For the moment, we only wish to emphasize that it is essential that these fish exert exquisite control over their electric discharge organs, and this control is provided by the central nervous system.

We usually think of sensory systems in terms of receptors and neural circuits that convey information from the outside world into the brain. Conversely, we think of motor systems, those that control movements, as conveying information from the brain to the muscles. It seems a fairly simple relationship: sensory systems provide inputs to the brain, motor systems convey outputs from the brain that produce movements.

This simple dichotomy is misleading, however, for it overlooks the fact that movements alter the nature of incoming sensory information, and that incoming information in turn alters movements. Perceptual processing really consists of reciprocal relationships between sensory systems and motor systems. Although it may seem counterintuitive at first, many sensory systems cannot be understood without considering the importance of movement. For instance, vision depends on movements of the eyes. Even when asked to maintain your gaze on this plus sign,

+

your eyes make small movements that are detectable only with special equipment. These eye movements prevent the + from being truly stationary on your retina: no matter how hard you try to keep your eyes still, the image of that little + is wandering on your retinas—jiggling across the array of your photoreceptors in an essentially random fashion.

You might think that such small, erratic movements of retinal images is harmful to good vision. After all, it would certainly be difficult to read if your book were to start jumping around in unpredictable ways. It turns out, however, that these erratic motions of the eyes are not at all deleterious to vision. In fact, they are essential for vision. To demonstrate this, we use specialized equipment that moves the + as you look at it, in exact correspondence with your miniature eye movements. This procedure, which can actually be performed in a laboratory setting, is called retinal stablization. Under conditions of retinal stabilization, the image of the + remains perfectly stationary on your retina. When that happens, the + will actually diappear! Sustained vision actually requires these little random motions of your eye. (See figure 15.2.)

Figure 15.2
A demonstration of the effects of eye movements on maintaining the visibility of objects. Because the spot on the left is so blurred and faint, the tiny eye movements that ordinarily maintain visibility are not sufficient. If you look steadily at the center +, and are patient, the gray spot will disappear. In order to experience this effect, you must fix your gaze very steadily on the center +, but pay attention to the appearance of the spot. When it fades, shift your gaze to the rightmost +. The spot suddenly reappears! If you keep your eyes steady, the spot will again disappear, and shifting your gaze back to the center plus will restore its visibility. If the edges of the spot were sharp rather than blurred, retinal stabilization would be needed to produce the same effect.

The sense of touch provides another example. People are much better at identifying a complex object when they can actively feel it than when it is passively pressed against their hand. Active touch provides input that guides subsequent manipulations of the fingers, which in turn provide new sensory data. The interaction between touch and movement provides a much richer source of information than passive touch alone.

Similar principles apply to electroreception. The "novelty response" is but one example. When a pulse-type electric fish detects a novel object, it increases the rate of its EOD, actively exploring the newly detected object in a manner analogous to the way you would actively touch an unseen object by manipulating it with your hands. Clearly, this novelty response requires that sensory information be conveyed from relevant electroreceptors to the electromotor control system. That information must then be used to effect the appropriate change in ongoing EOD.

The neural control of the electric organ is often regarded as a model system of the neural control of motor functions more generally. In comparison with something like human motor coordination during the performance of a Chopin etude, it is a fairly simple system to study. Changes in EOD frequency are obviously much easier to analyze than the movements that occur during a piano performance or an expert golf swing (if that were not true, the marketing of golf magazines, books, and videos would not be necessary). From the scientific perspective, a good model system is one that embodies design characteristics that are important features of more complicated systems, but remains simple enough to understand using current scientific methods. The electromotor control system illustrates some organizational principles that are common features of all motor systems, yet is relatively simple in its organization.

Hierarchical organization is one of those general principles. Motor systems are organized into different levels of control, just like the electric organ control system. Lowest in the hierarchy are the neurons that innervate muscles (or in electric fish, the modified muscle cells that comprise the electric organ). Higher levels in the system exert their effects through control of the lower levels. These higher levels also are usually associated with more complicated functions that require specific patterns and sequences of activity among lower-level elements (think of the patterns and timing of finger movements in a skilled pianist, or the timing required for synchronous discharge of the cells in the electric organ of the fish). The highest levels of control are often "executive" in nature: they command a certain movement or pattern of movements, but, like the CEO of a large corporation, are not concerned with the details of how these commands are executed.

In the fish, the so-called electromotor neurons represent the lowest level of neural control over the electric organ discharge. These neurons are found in the fish's spinal cord, and their axons leave the spinal cord and contact the electrocytes, the cells that constitute the electric organ. Some interesting anatomical adaptations are used by the spinal electromotor cells to ensure synchronous activity among the electrocytes. The spinal electromotor neurons all discharge synchronously, but it is not a trivial matter to achieve the required temporal precision. It is achieved in part by specialized electrical connections between the spinal cells. These electrical synapses between the electromotor cells help ensure that when one is activated, that activity quickly spreads among them all, so that they act as a unit. But synchronous activity among these cells is not enough to ensure synchronous discharge of the electrocytes.

Why not? Many electric organs are fairly long, so the path from spinal motor neurons to the electric organ varies in length: some neurons have a shorter path and others have a longer path. The importance of these differences in path length is that nerve impulses are conducted at a finite velocity. The range of conduction velocities in these fish range from about

10 to 40 m/sec (22–88 m/hr). So, even if all the electromotor neurons emit a nerve impulse at the same time, the impulses would not arrive at all the electrocytes at the same time—the more distant electrocytes would be activated after those that are closer. There are two obvious ways in which these arrival times could be equated. One solution is simply to equate the path lengths: the axons that innervate the nearby electrocytes can take a circuitous route to their electrocyte targets, and those that innervate more distant electrocytes can take a more direct route. The other is to systematically vary the conduction velocities of the axons such that the shorter axons conduct more slowly and the longer axons conduct more quickly. This can be accomplished by carefully varying the size of the axon fibers, since large fibers conduct impulses more quickly than do smaller fibers. Both strategies seem to have been adopted, in different species (Bennett, 1971).

This sort of arrangement—smaller fibers innervating nearby electrocytes and larger fibers innervating more distant ones—requires a remarkable degree of specificity in the connections between the electromotor neurons and the electrocytes. How do the large fibers "know" to bypass the nearer electrocytes in favor of those that are more distant? The question of the specificity of connections between neurons is one of the great mysteries of brain science. It is a ubiquitous issue, for neuronal circuits typically display amazingly specific patterns of connectivity. How do these intricate wiring diagrams develop? Are they genetically preprogrammed? To what extent do they depend on the experience of the individual? In general, it appears that the patterns of nerve cell connections depend on both genetic factors and experiential ones. In many systems, it appears that "target cells" release chemical compounds that help cells "recognize" when they have made the appropriate synaptic contacts. Often, there is an "exuberance" of connections made early in development, and, as development proceeds, many of these extra connections are pruned away. Another feature in the development of organisms is that many more neurons are produced than will ultimately survive. There appears to be a competition between developing neurons, and only the winners survive into adulthood.

The electromotor control system provides a good illustration of these types of specificities. It is notable in this context that many electric fish

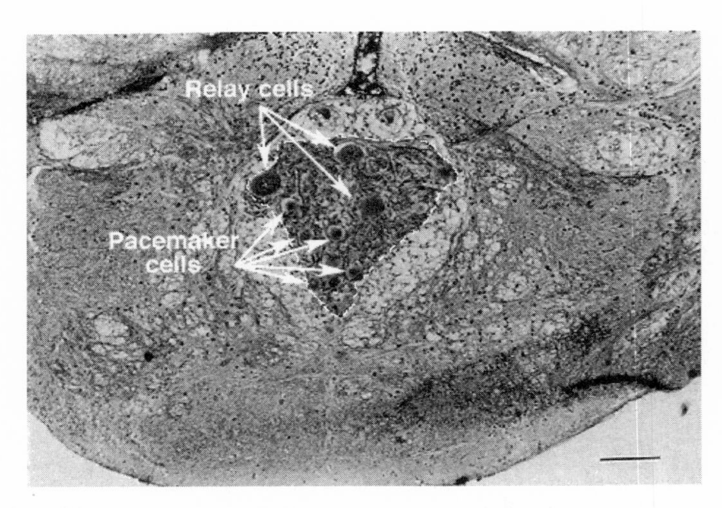

Figure 15.3
The pacemaker nucleus of one South American variety of weakly electric fish. These neurons are located in the medulla of the fish's brain. The larger cells project to the spinal cord, and control the spinal electromotor neurons. The smaller cells control the larger cells, illustrating the principle of hierarchical control.

have a larval electric organ that disappears during development and is replaced by an adult organ. In many cases, the larval and adult organs are found in different locations on the body. Each organ produces an electric discharge, and each receives an innervation. Clearly, dramatic changes in the electromotor control circuits must take place as the larval organ disappears and is replaced by the adult version.

Since the electrocytes are controlled by electromotor neurons of the spinal cord, we naturally wonder what controls the electromotor neurons. A major source of this next level of control in the hierarchy comes from a specialized group of neurons in a portion of the brain called the medulla. All vertebrates have a medulla, and the medullae of all vertebrates contain neurons that control spinal motor neurons. This medullary relay nucleus represents the second level of the hierarchy. The medullary relay cells control the spinal motor neurons, which in turn control the synchronous activity of the electrocytes that produce the EOD. Similar hierarchies exist in all vertebrates, with the obvious difference that spinal motor neurons ordinarily control muscles rather than electrocytes.

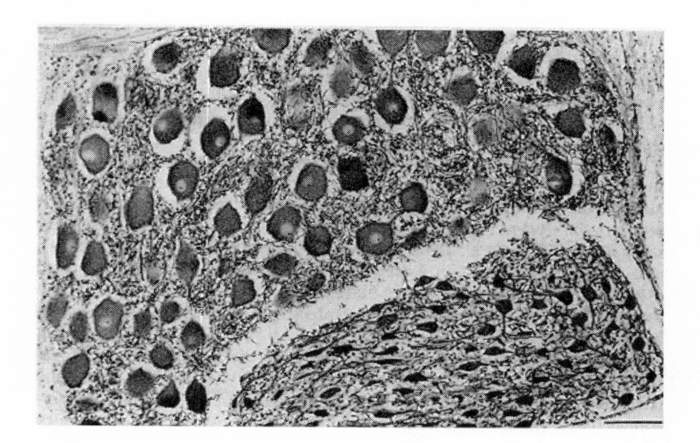

Figure 15.4
A photograph showing the pacemaker nucleus of species in which the relay cells (larger neurons on the top) and the pacemaker cells (smaller neurons on the bottom) are anatomically segregated.

At the top of the electromotor "chain of command" lie the pacemaker cells. In many species, these cells are aggregated into a distinct group of neurons known as the pacemaker nucleus. The precise timing of EODs is achieved by these cells, which serve as the command center for the electromotor system as a whole. As you might expect in any hierarchical organization, there are often fewer neurons at each successively higher level of the hierarchy. In most armies, the privates are greatest in number, followed by the lieutenants. Generals are relatively rare. In one species, it was estimated that 500 or so electrocytes were innervated by 250 spinal electromotor neurons, which in turn were controlled by only 25 pacemaker cells.

What army would have more generals than lieutenants? While this might not be good military organization, it seems advantageous to some species of electric fish, which have more pacemaker than relay cells. Although we do not yet understand the advantage of this reversal in numbers, we presume that there is one. In any case, control of the EOD is implemented by controlling the pacemaker nucleus. What kinds of factors control the pacemaker? Social influences are one of the most important sources of control over the EOD.

16

Electroreception in the Social Context: Better Living through Electricity

The population of electric fish can become quite dense, and when the fish get sufficiently close to one another, we might expect their signals to cause interference. Similar issues arose with bat sonar. Immunity from interference from other bats was achieved in part through the use of the combination cells in the bat's auditory system. Combination cells respond to a specific combination of the fundamental frequency of the bat's call and one of the higher harmonics. Since the fundamental is transmitted primarily through the bat's head, bats will not hear the fundamentals of other bats in most circumstances. In this way, the bats are naturally most sensitive to their own call-echo interactions, thus minimizing the potential interferece between individuals.

How do electric fish avoid interference from the fields of their neighbors? In the jargon of radar operators, how do the fish prevent jamming of their signals by other electric fish?

Electric fish use a variety of techniques to avoid being confused by the myriad signals that intermingle in densely populated waters. Differences in the EOD waveform of different species helps to minimize interference between species. Lissman compared the electrical signals of *Gymnarchus* with those of several other African species that live in the same habitat. He also compared these EOD waveforms with those of some South American species. Lissman noted that the EOD of each species differed in the shape of the individual electric pulse and the rates at which it emitted these pulses. It appeared that each species had a unique electric

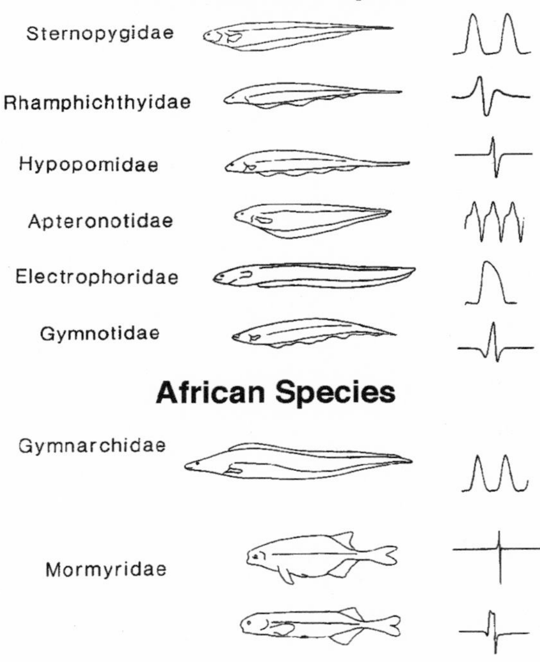

Figure 16.1
Different species of electric fish differ in the waveforms produced by their electric organs. These differences, which to a large extent depend upon the anatomy of the electric organ in each species, help to minimize interference between different species.

"signature." The shapes of the individual discharges of several different species of electric fish are illustrated in figure 16.1. The EOD of each species is different.

Pictured in figures 16.2 and 16.3 are examples of the EOD waveforms of several species. To the right of each waveform is the corresponding amplitude spectrum for each pattern of electrical discharge. These graphs illustrate the different sine wave components that, if added in the proper phase relationships, sum to form the observed EOD of each animal. Frequency spectra such as these are often used to analyze signals because they conveniently represent the component frequencies that dominate the waveform. This is useful information because if two signals differ in their

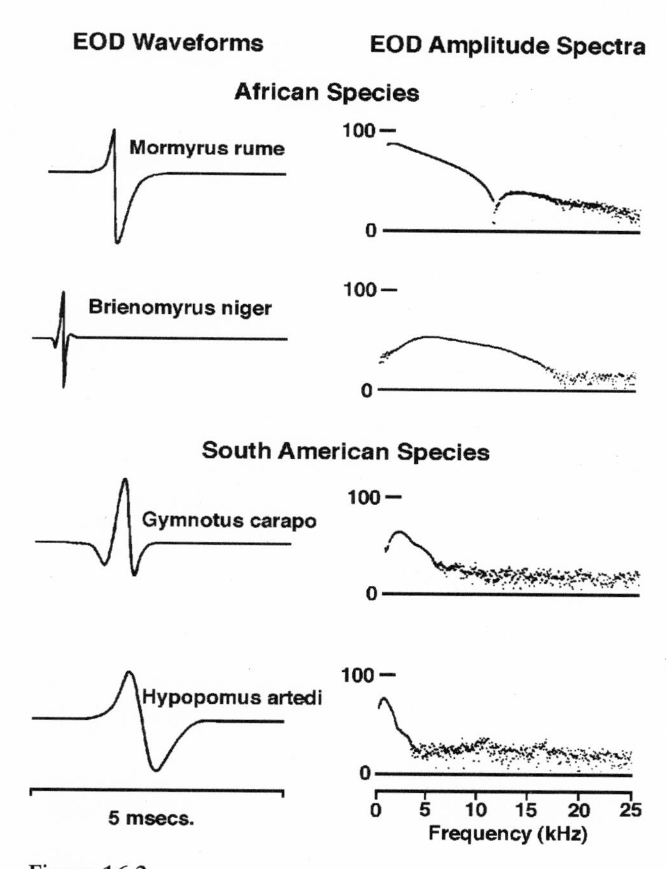

Figure 16.2
EOD waveforms and amplitude spectra for several varieties of pulse-format electroreceptive fish. Interference between members of different species could be minimized if each fish's electroreceptors were tuned to a frequency (or pattern of frequencies) unique to its own EOD.

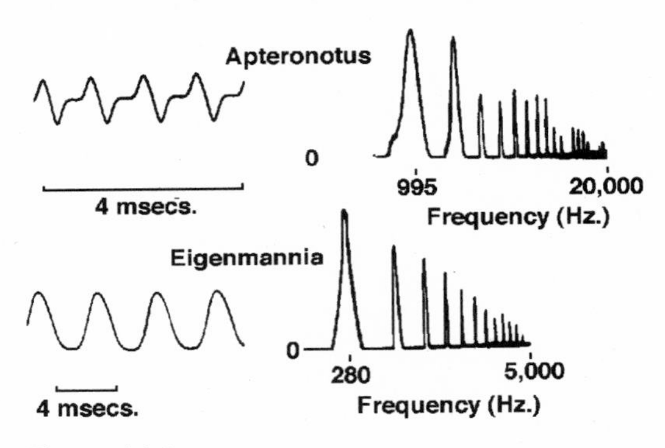

Figure 16.3
Waveforms and amplitude spectra of two South American species of wave-format electroreceptive fish. The frequency of *Apteronotus* is about four times greater than that of *Eigenmannia.* The amplitude spectra in this case are similar, in that each has a fundamental and a series of harmonics, but the fundamental frequencies are quite different. Jamming can be thus be avoided if each fish has tuberous receptors that are tuned to its own fundamental frequency. In that case, tuberous receptors of *Aperonotus* would not respond to *Eigenmannia*'s EOD, and vice versa.

frequency content, an engineer can readily design a device that will detect one signal and not the other. In the context of our electric fish, if the signals of different species differ, the resulting differences in the amplitude spectra could provide the key to reducing interference among individuals of different species: they simply would evolve receptors that are selectively sensitive to the frequencies that dominate their own EOD. Comparisons between the frequency spectrum of the EOD and the sensitivity of the tuberous organs has repeatedly revealed a very close correspondence between them. This means is that each species' tuberous organs are electrically tuned to match its own EODs.

It is not entirely clear how this tuning is accomplished. An electrical engineer could produce a device that mimics the tuning of a tuberous receptor by using a combination of resistors, capacitors, diodes, and the like. An example of such an "equivalent circuit" is illustrated in figure 16.4. It is the exact arrangement and values of the different components that produce the desired frequency selectivity. Of course, the tuberous

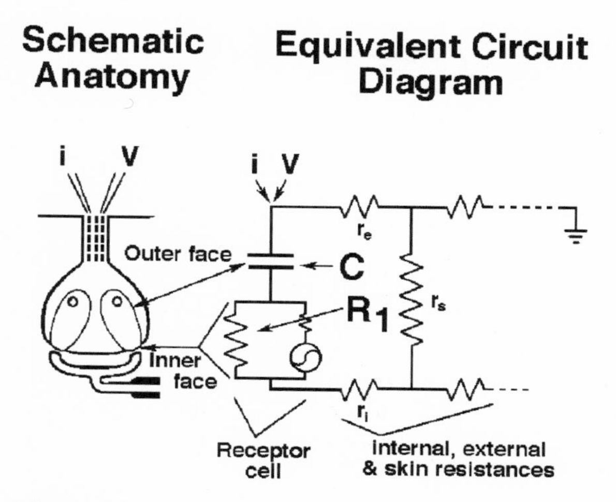

Figure 16.4
Equivalent circuit of a tuberous receptor. The combination of a capacitor (C) and resistor (R_1) acts to remove low-frequency inputs. Variation in the exact values of R_1 and C determine the frequency tuning of the receptor. This is accomplished in the tuberous receptor by means of anatomical specializations.

receptors of electric fish are not made of resistors, capacitors, batteries, and diodes. The tuning characteristics of electroreceptors depend upon the anatomical arrangements of the different cells—the packing of the cells that line the canal, the distances between their adjacent membranes, the presence or absence of ion pores that can regulate the flow of ionic currents across different membrane surfaces. All these factors play a role in determining the specific properties of each species' tuberous receptors.

So the exquisite match between the electric organ and the tuberous receptors provides a system in which the tuberous organs can be used to detect the self-produced electric field without interference from other species of electric fish. It's as if nature has established something like the Federal Communications Commission, and has allotted each species a broadcast frequency in the same way the FCC allots each station a broadcast frequency (see figure 16.5). Both are intended to prevent unwanted interference between neighbors.

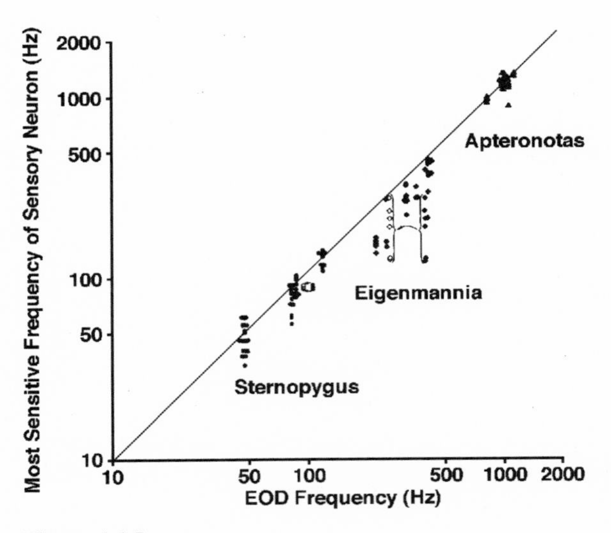

Figure 16.5
Illustration of the close match between the fundamental frequency of the EOD of several South American wave species of electroreceptive fish, and the tuning characteristics of the sensory fibers that innervate the tuberous receptors. If the match were perfect, all the data points would fall on the diagonal line.

The precise matching of the EOD and the tuberous organs does not stop there, however. Like human voices, individual fish within a species have slightly different EODs, and their tuberous organs are tuned to these slight individual differences. Males have EODs different from females (like humans, the males tend to broadcast at lower frequencies). Dominant males have the lowest frequency components of all. The tuning characteristics of each individual's tuberous electroreceptors match their individual discharge frequency. In fact, just like the voice of a teenage boy, the EOD frequencies of many electric fish change during development. These maturational changes are accompanied by carefully matched changes in the tuning characteristics of the electroreceptors. The match between the frequency characteristics of an individual's EOD and the frequency tuning of the tuberous electroreceptors make the receptors especially sensitive to the fish's own EOD, and therefore helps prevent jamming of an individual's electroreceptive system by the EODs of other fish, especially those of other species.

These fish might not want their electroreceptive systems to be completely insensitive to signals from other species, however. Indeed, a complete insensitivity to the signals of other fish would render electroreception useless as a medium for communication between individuals. These fish do indeed use their electric discharges to communicate with others, so their electrosensory systems not only must provide them with the noise immunity needed for active electroreception of the environment, but they must also, at the same time, *preserve* a sensitivity to the discharge of others, so that they can stake out territorial waters, establish social hierarchies, and find and attract mates. The fish have evolved some remarkable mechanisms for meeting these competing needs.

For instance, an African family of electric fish called mormyrids have three different types of electroreceptors. Like most aquatic animals, they have ampullary receptors. In addition, they possess tuberous receptors of two different types. One type is designed to sense the self-produced electric field generated by the fish's EOD. The other type, the *knollenorgan,* is specifically designed to detect the discharges of *other members of the same species.* Mormyrids thus solve the problem of separating self-produced EODs from those of other individuals of the same species by employing different receptor systems that are selectively sensitive to these two different sources of electric signals. The sensory nerves that convey information from the knollenorgan have different connections within the fish's brain than the receptors that detect the animal's own EOD, so that different neural circuits are involved in the active electroreception sense and what we might call the electroreceptive communication system. In this sense, mormyrids have three different electroreceptive systems. Passive electroreception is mediated by the ampullary organs. Active electroreception is mediated by one class of tuberous receptor. And social communication is mediated by another class of tuberous receptor (the knollenorgan). The fish use some clever techiques of signal processing to achieve the receptor selectivity required of these three functionally distinct sytems.

In order to appreciate the complexity of the problem, consider the plight of the genus *Mormyrus*, an example of which is illustrated in figure 16.6. Like many of the African varieties, this is a pulse species. Each individual pulse is brief in duration. The exact shape of the wave is

Gnathonemus tamandua

Figure 16.6
Example of the genus *Mormyrus,* an African variety of electroreceptive fish.

specific for the species and differs in males and females. In the behaving animal, the rate at which these pulses are emitted varies dramatically, depending on the situation. We've seen already that these fish produce a novelty response: they increase the rate of the EOD when they detect a new object in the environment. They also vary the rate in social situations, such as when they encounter another individual in their territory, or during courtship and mating.

Unlike wave species, whose EOD looks almost like a single sine wave, the EOD waveform of *Mormyrus* contains a wide range of frequencies. This is because the individual pulses are so brief in duration. This broad frequency spectrum of the EOD has an important consequence for the fish: the frequency tuning of the electroreceptors will not automatically filter out the EOD. Thus, even though the ampullary receptors are tuned to low frequencies, they will respond to the fish's EOD because the EOD contains low- as well as high-frequency components. In *Mormyrus*, all three receptor types respond to the fish's EOD, and to the EOD of other members of the genus. So if all three receptors respond to these EODs, how can the fish selectively use the ampullary receptors for passive electroreception, the knollenorgan for social communication, and the third receptor type to specifically analyze the self-produced fields needed for active electroreception?

In order to produce the required selectivity in their three types of electroreceptors, the mormyrids use a method of signal cancellation that is commonly found in many sensory systems, including perhaps the most remarkable sensory system of all: the human sense of vision. The general idea of the technique works like this. The goal is to preserve the sensitivity of some receptors to the individual's own EOD, while at the same time eliminating the response of other receptors. We have seen that the EOD is initiated by neural signals that occur in the fish's brain. What would be the effect of taking those electromotor command signals and, in addition to sending them to electromotor neurons that will initiate the EOD, sending a copy of the electromotor command to sensory centers in the fish's brain? Each time an EOD command is issued, the sensory processing system is informed of the upcoming EOD. That information could be used to inhibit sensory cells, rendering them temporarily insensitive to electric signals. This idea that motor commands can be used in sensory systems to negate the effects of movements (or, in our case, electric discharges) that would otherwise interfere with sensory processing is called "efference copy." Efferent systems are essentially motor systems, so the term "efference copy" effectively captures the idea that a copy of the motor command is being conveyed or fed back into the sensory system.

Mormyrids make use of these efference copy signals in several different ways, depending on the specific needs of each receptor system. For instance, the response of the ampullary receptors to each EOD is fairly long (up to about 0.1 second). The fish can generate multiple pulses in each second. If the efferent feedback signal were simply to inhibit cells that normally would respond to electrical changes detected by ampullary receptors, the ampullary pathways would be "shut down" for much of the time. This is obviously an undesirable solution: just think how difficult it would be to see if eye blinks lasted as long as a second. If we blinked every 5 seconds, we would be completely unable to see 20% of the time. So simply using a copy of the electromotor command signal to shut down the ampullary organs during each electric discharge will not work as the fish would like. Something more elaborate is needed.

What appears to occur in the ampullary processing pathway is that the efference copy signal that is fed back to the sensory system is an inverted copy of the ampullary response to the fish's EOD. That is, the signal follows the same time course as the ampullary response, but it has the opposite sign. The signals from the ampullary receptors are then summed with the efference copy signal. Since they are mirror images of one another, this summing operation effectively removes the EOD contribution of the ampullary signal, but leaves any residual signals unaffected. Using this elegant strategy, the ampullary processing centers can remove the EOD contribution, and the only thing that remains is the low-frequency signals produced by other sources, such as the gill movements of other fish. In this way, passive electroreception proceeds without interference from the fish's own electric discharge.

Efference copy is also operating within the neural pathways specialized for processing signals from the knollenorgan. These receptors respond to each EOD pulse by producing a single impulse of their own. The knollenorgans are quite sensitive to the species-specific EOD, and their response to an EOD pulse does not vary with the strength of the EOD signal. This means that the receptors will respond in the same way to the discharge of another fish, no matter how far away the other fish happens to be (so long as it's close enough to produce a detectable EOD). In order for the knollenorgan to serve as a receptor for the detection of the EODs of other members of the same species, the receptor should be tuned to the frequencies that are characteristic of that species, and it must also be selectively responsive to the discharge of other individuals. In the ideal case, the knollenorgans should not respond to self-produced EODs. What is needed in this case, then, is a mechanism that will selectively inhibit the knollenorgan response to the self-produced EOD, but not to those of other fish.

Just such a mechanism was described in 1969 by Bennett and Steinbach, who were working at the Albert Einstein College of Medicine in New York City. They took advantage of a drug that blocks the communication between the motor neurons and the modified muscle cells that comprise the electric organ. The drug, curare, is the substance that native peoples of the Amazon rain forest use on the tips of their arrows. When one of these arrows hits a living target, the curare passes into the bloodstream, and blocks transmission from neurons to muscles. Paralysis results, and the victim dies because the diaphragm is paralyzed along with all the other muscles—the actual cause of death is suffocation. Heart muscle is not affected. The victim is completely aware of all that occurs right until the end.

In any case, curare applied to electric fish will prevent EOD, but not the neural commands that precede it. So if the fish is artificially respirated, the experimenters are left with a fish that tries to produce an EOD, but can't. The critical feature of the experiment is that the EOD is disrupted at the last possible moment—at the connection between the electromotor neuron and the electric organ. When the fish attempts to produce an EOD, the efference copy—if it actually exists—should still be issued and, according to the theory, should still influence the sensitivity of the tuberous receptor. Attempts to produce the EOD can be monitored by recording electrodes placed in the electromotor command center of the medulla. Bennett and Steinbach recorded sensory activity in neural centers devoted to processing inputs from the knollenorgan while they simultaneously recorded electromotor command signals in the electromotor pacemaker nucleus. In addition, it was necessary to play EOD signals out of electrodes, in the manner first employed by Lissman. This way, the sensory response to a prerecorded EOD could be recorded in the absence of any interference from the fish's EOD. What is more, the sensory responses to an EOD can be obtained at the moment the fish is trying to produce a discharge of its own, and compared with times when the EOD command was not being issued.

Bennett and Steinbach observed that the sensory neurons would often respond to these synthetic EODs, but not always. In particular, each time the pacemaker nucleus generated an electromotor command signal, the sensory neurons would become completely insensitive to the artificial EODs. This inhibitory effect was very brief in duration; it only lasted 0.0015–0.0020 second after the peak of the electromotor command signal. That is just enough to specifically disable the knollenorgans during self-produced EODs. At all other times, the knollenorgan system remains sensitive to EODs, but these will always be the EODs of other fish. It's as if the fish either can "talk" or it can "listen"; but like us, it can't do both at the same time.

So efference copy is the method used by pulse-type mormyrids to distinguish between their own EOD, the EOD of nearby members of the same species, and the low-frequency electrical signals produced by all fish. But efference copy is not employed by all weakly electric fish; for unknown reasons, the South American varieties do not appear to use this strategy.

What we have discovered so far is a mechanism that enables mormyrids to process their own EODs using one system of tuberous receptors, and process the EODs of other fish using a different class of tuberous receptors (the knollenorgans). Disturbances in the fields produced by the fish's own EOD is used for electrolocation, the process of detecting and localizing objects by means of the electric sense. Processing the discharges of other fish is important in social communication, including mating. Note that the efference copy does not, in and of itself, protect the fish from interference from other fish. For example, if two mormyrids were to come close to one another, their electrical fields would sum together in complex ways. If the two fish were producing EODs at the same moment in time, the efference copy would render each invisible to the other.

The joint requirements of electrolocation and social communication would be best served if these two fish could find a way to ensure that their EODs are never produced at the same moment in time. Mormyrids achieve this by "echoing" the EODs of nearby members of the same species. That is, when the knollenorgan system detects the discharge of another individual, the fish "echoes" that with an EOD of its own. The echo response follows the neighbor's EOD by 0.010–0.012 second. This delay is appropriate because successive EODs of individual mormyrids are very unlikely to occur during that interval. The echo response thus serves to ensure that neighboring fish do not produce simultaneous electric discharges, and therefore do not interfere with each other's electrolocation systems.

The problem of interference between the signals of neighboring electric fish is ever present, and while the echoing response works well for pulse species, it is not a feasible solution for those species that produce wave discharges. These so-called wave species do not produce a sequence of short pulses separated by longer intervals of electrical silence. The electric discharge of wave species is more like a sine wave: a continual fluctuation in voltage over time. Electrolocation in these animals depends on sensing changes in the amplitude and timing (technically called phase changes) that are produced by other objects in the water. The problem is that the signals of neighboring fish will sum with one another, and the resulting sums vary in amplitude and phase in a way that mimics other objects. How can the fish avoid mistaking the combination of electric signals from itself and other fish for inanimate obstacles or potential meals?

The outlines for an answer to this problem began to appear in 1963, when Watanabe and Takeda discovered that the South American waveform genus called *Eigenmannia* shifts the frequency of its EOD when confronted with an interfering signal of similar frequency. If the second signal is slightly higher than *Eigenmannia*'s ongoing EOD frequency, the fish lowers the frequency of its EOD. If the second signal is lower than the EOD frequency, *Eigenmannia* reponds by increasing the frequency of its EOD. The change always increases the difference between the frequency of the experimental signal and the fish's EOD waveform. Because the purpose of this behavior appeared to be the avoidance of interference from the second signal, these changes in EOD frequency were called the jamming avoidance response (JAR) by Theodore H. Bullock and Walter Heiligenberg of the Scripps Institute of Oceanography. Bullock, Heiligenberg and others began an intensive investigation of the JAR in the early 1970s, and their work has led to a remarkably detailed understanding of this behavior and the neural circuits that control it. So we turn now to take an appreciative look at the JAR.

The JAR is a marvel of biological engineering. When two *Eigenmannia* encounter one another, the changes in their two EOD frequencies always increase the difference between them. That is, the individual with the lower EOD frequency always lowers its EOD frequency and the fish with the higher frequency always increases its frequency. This means that each fish must be able to sense the very slight differences in the frequencies of the two signals, decide whether the frequency of the second signal is higher or lower than its own, and make the appropriate change in its EOD frequency. To me, it calls to mind those situations we have probably all encountered when you see someone walking toward you in a corridor or on a sidewalk. Being polite, you decide to alter your direction slightly, to veer to the left or right so as to give the other pedestrian a clear path. You go left, and at the same moment, the person walking toward you decides to go right. You're still on a collision course. Then you go right, the other person goes left. Still about to collide. You both laugh, decide how you are going to manage this seemingly simple maneuver, and then each go your own way. *Eigenmannia* never have this problem. Their adjustments always increase the frequency difference in the two EODs, and thus each always minimizes the interference of the other. They are obviously very adept not only at determining a small difference in the frequencies, but also at establishing which fish has the lower frequency.

Perhaps an illustration will heighten our appreciation for how hard a task this really is. Like all wave-type species, *Eigenmannia*'s EOD is closely approximated by a sine wave. Let waveform EOD_1 in figure 16.7 represent the EOD of one fish, and waveform EOD_2 represent a distant but detectable neighbor. Since Fish #2 is relatively far away, the strength of the EOD_2 signal near #1 is smaller than 1's own signal. The electric fields created by both fish will be summed in the water. In some part, a peak in 1's signal will be added to a trough in 2's, and in other places, the peaks will add. As a result, when the two signals are added, the sum is an amplitude-modulated signal. Such a sum is illustrated for two cases; one in which EOD_1 has a slightly lower frequency than EOD_2 and one in which the frequency of EOD_1 is slightly higher than EOD_2. So the proximity of another fish produces an amplitude-modulated field, and that is a problem for the fish, since objects in the water can disturb the electrical field in a similar manner.

The rate or frequency of the amplitude modulation is determined by the difference in the frequencies of EOD_1 and EOD_2. If the difference is very small, the frequency of amplitude modulation is low. Increasing the difference between the frequencies will increase the rate of amplitude modulation in the sum, and when that rate becomes sufficiently high, it no longer mimics the effects of objects in the water.

Signals EOD_1 and EOD_2 in figure 16.7A are not exactly the same frequency: EOD_1 is slightly slower than EOD_2. The appropriate JAR in this case would be for fish 1 to decrease and fish 2 to increase their EOD frequencies, which is what they do. But how do the fish know which frequency is higher? They clearly must derive that information from the amplitude-modulated patterns produced when the two signals are summed in the water. Compare the two summed EOD signals in figure 16.7. They are both sums of two sine waves. In the left waveform, the frequency of EOD_1 is slightly lower than that of EOD_2. The situation is

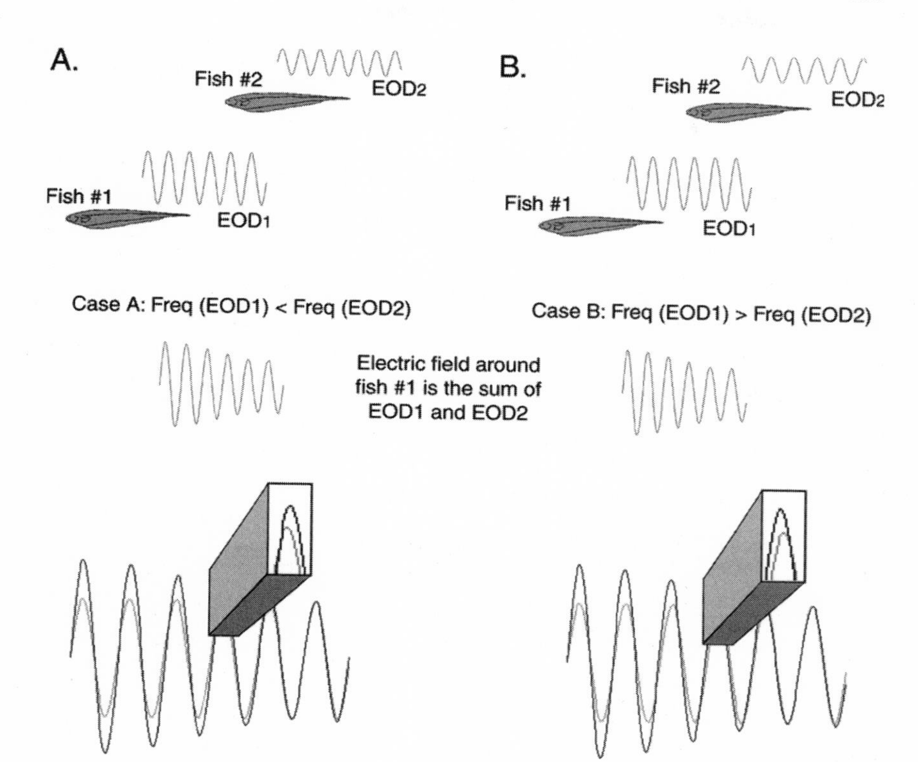

Figure 16.7
This is an illustration of the subtlety of the cues that wave-species of electric fish use during their performance of the jamming avoidance response. In A, the EOD frequency of fish #1 is lower than fish #2, so the appropriate response of fish #1 is to lower its EOD frequency. In B, the EOD frequency of fish #1 is higher than fish #2. In this case, fish #1 should increase its EOD frequency. The electric field surrounding #1 is the sum of the EODs of both fish. As indicated in the middle panels, these fields are very similar in both situations, yet the fish invariably alter their EODs appropriately. The bottom panel illustrates the summed waveforms (dark traces) and the waveworm of fish #1 produced in isolation. Comparisons of the two show there are slight variations in the timing of the peaks in the sum relative to the fish's own EOD, and those variations differ in A and B. The fish can detect these phase-shifts by comparing the electric field as sensed by tuberous receptors and an efference copy of the individual's EOD. It can then use that information in determining whether the appropriate response is an increase or decrease in EOD frequency.

reversed in the right waveform. *Eigenmannia* is able to sense the differences in waveforms such as these, and correctly identify which of the two EODs is higher.

A closer examination of the two waveforms reveals that the peaks of the summed waveform occur at different times relative to the EOD of fish 1 occurring in isolation. (See insets at bottom of figure 16.7.) These shifts in the position of the peaks relative to one another are called phase shifts. Although not illustrated in figure 16.7, we also should note that the peak of the combination is in some places greater than the peak of EOD_1, and in some places it is lower. So the sum of these two simulated EOD signals has a pattern of phase shifts and amplitude changes relative to what the fish's isolated field produces in the absence of a neighboring fish. The patterns of these phase shifts are different when the neighbor's EOD frequency is higher and when it is lower, and this difference provides the information that allows the fish to alter its EOD frequency in the appropriate way. Of course, it senses these differences using its tuberous electroreceptors, which are distributed along the length of the animal and can therefore sample the complex waveform produced by the presence of other fish at different spatial locations. Since the phase shifts produced by a neighboring fish are different when the fish has a lower EOD frequency than when it has a higher one, *Eigenmannia* can correctly determine the sign of the difference frequency and change its own EOD accordingly.

The JAR is but one aspect of electroreception that is influenced by social contexts. Many others have been identified. In morymrids, which are primarily pulse-type species, changes in the time intervals between electric discharges have been shown to convey a remarkably rich set of social messages. These range from aggressive signals, through warnings of an impending attack, to courtship and mating rituals. Indeed, mating in these fish is accompanied by a sequence of electrical overtures, referred to as "courting and sparking" in an article written by Hagedorn and Heiligenberg (1985). Here is a bona fide case of "animal electricity"!

Scientific interest in electric fish is based primarily on the pure goal of understanding more about the world around us. Electroreception appears to be a primordial sensory system; one that apparently has been employed since vertebrates first evolved on the planet. As might be expected from such an ancient sensory modality, the neurobiology of electroreception has many similarities with other sensory systems, and has provided many widely applicable insights, such as the intimate relationship between motor and sensory systems epitomized by the JAR. Of course, it is always nice if practical benefits are derived from such basic scientific research.

We know the ancients believed that the strongly electric species had some medicinal value, but certainly we have more easily implemented ways of using electrical signals in medicine today. One practical application of our knowledge of electroreception that readily comes to mind is the possibility that we might be able to develop a device that protects people from shark attacks. The Natal Sharks Board, an organization based in South Africa, has been working on just such a device. They call it the Shark POD (protective oceanic device). The Natal Sharks Board has a page on the World Wide Web in which it describes the POD as the world's first effective electronic shark repellent. It is intended for use by scuba divers, and is claimed to be harmless to both people and the sharks. It produces an electrical field that repels sharks that come within 3 meters of the POD. The makers of the POD hope that future developments will lead to "electrical fences" that will keep dangerous sharks out of bathing areas, for in certain parts of the world, sharks are all too frequent visitors to local beaches.

IV

The Scents of Attraction

17

Chemical Communication via Pheromones

A World Wide Web page (‹http://www.athena-inst.com/10x.html›, visited on 8/20/97) promises answers to these and other timeless questions:

"Why do women tend to end up end up on 'his' side of the bed?"

"Why does she wear his shirt or robe when he's away?"

"Why does a child's security blanket or teddy bear lose its appeal after it is washed?"

"Why do blondes have more fun?"

The answer proposed in each case was the same: "pheromones." Now some skepticism might well be appropriate here. First, we might question the premises themselves. Do women really tend to migrate to "their man's" side of the bed? Do they really like to wear "their man's" clothing? Do blondes actually have more fun? So far as I am aware, there is little in the way of scientific evidence for any of these suppositions. In my own experience (as a parent; I can't remember my childhood in that much detail), the most valid of these assumptions might be that a baby's security blanket really does lose its appeal after washing.

But let's set any of our doubts about the questions aside for a moment, and focus on the suggested answer, pheromones. Pheromones are chemical messengers. In many species, they play an essential role in communication between individuals. Some pheromones communicate fear or danger to all the members of a group. The release of such "alarm" pheromones into the air (or water, for certain fish also use alarm pheromones) leads to immediate dispersal of a group or to defensive behaviors. For instance, glandular secretions of endangered worker ants are readily detected by other ants within the colony, and they mobilize the colony to defend the nest. At least four different alarm pheromones are released by ants, and ant colonies respond to the experimental introduction of these substances as they would to a real emergency.

Another class of chemicals is called "dispersion" pheromones. They act to maintain an optimal separation between individuals and are used, for example, to mark individual territories. Who hasn't noticed that peculiar canine habit of depositing small amounts of urine on trees, fence posts, and the proverbial fire hydrant? Many canine and feline species mark the borders of their territories with urine. The urine contains pheromones that are detected by other members of the same species, and indicate to the recipient that he or she is trespassing into another's territory. Not all dispersal pheromones are released in urine, however; rabbits release them from a gland located in their chin, and many other species release them from glands located near the anus.

One type of dispersal pheromone signals death in certain species of social insects. Chemicals released by a dead individual's body are detected by others in the colony and lead to the immediate removal of the deceased to a "refuse center" outside the nest. If these chemicals are placed on a live ant, it is unceremoniously transported to the refuse center. When it returns, it is again escorted to the ant cemetery. These premature interments continue until the scent has worn off.

Other pheromones serve to attract individuals. Such "aggregation" pheromones can perform important functions like attracting individuals to a rich food source or to an ideal site for nesting or colonization. Notable among such pheromones are the sex pheromones, and it is these chemicals that have drawn so much media attention. Indeed, several perfumes claim to contain human sex pheromones, which are said to exert powerful effects on members of the opposite sex. We will certainly want to evaluate the accuracy of those claims.

Sex pheromones can be released by both males and females. The females of many species release particular substances when they become sexually receptive. Often, female sex pheromones are highly volatile substances; they readily diffuse into the air, advertising the presence of

Figure 17.1
The elaborate antennae of the male silkworm moth are remarkably sensitive to the scent of a receptive female moth. As the saying goes, they can smell a female "a mile away."

the receptive female. The males of the species are equipped with sensory receptors (called chemoreceptors) that are exquisitely sensitive to these compounds. The presence of these sexual pheromones has astonishingly powerful effects on males of the same species. The silkworm moth is a particularly well documented case. In fact, bombykol, the sex pheromone of the female silkworm moth, was the first animal pheromone to be isolated and chemically identified. It took 500,000 females to produce just 15 milligrams of bombykol. The substance is released from a gland in the female's abdomen, and has a profound effect on the behavior of males. They follow the chemical trail left by the female, and when they find her, they mate.

Tracking the scent of the female requires a remarkably sensitive sensory apparatus. But male silkworm moths don't have a nose. They detect the bombykol by using their elaborate antennae (figure 17.1). Chemoreceptors in the antennae are specifically sensitive to bombykol: even a slight alteration in the chemical structure of bombykol renders the substance ineffective as a sexual attractant. The complicated branching structure of the antennae maximizes their surface area, so that many bombykol receptors can be employed in sampling the air. Although they are insensitive to many chemical "impostors," the bombykol receptors are exquisitely sensitive to the real thing: an individual receptor cell can

Figure 17.2
As this figure illustrates, male moths (top) tend to have much larger antennae than females.

signal the presence of a single molecule, and only 200 molecules can initiate a behavioral response (searching for the female). It is estimated that males can detect airborne bombykol at a concentration of 1 part bombykol per billion air molecules. Since it is the males that must find and pursue the females, males tend to have much larger antennae than the females (figure 17.2).

Being airborne, bombykol is dispersed by wind. Once the substance is detected, the males must follow a concentration gradient, constantly seeking higher concentrations of the pheromone until they find the source. Release of sex pheromones can attract males from a distance of several miles. I suspect the vast majority of women would find such a powerful allure completely undesirable—something we should bear in mind when evaluating claims for the actions of human pheromones.

Interestingly, although silkworm moths can see, vision has little to do with finding mates. Males will not approach a female placed in a glass jar (which permits vision but prevents the pheromone from reaching the male), but will approach a female that is confined to a mesh container that prevents vision but permits diffusion of the pheromones.

Male insects often produce pheromones that are essential to successful reproduction. The males of many butterfly species are equipped with "brush organs"—feathery organs consisting of many fine filaments that, like the antennae, dramatically increase their surface area. Once a male encounters a receptive female, it tends to fly above the female and bring its brush organ in close contact with the females' antennae. Pheromones released by the brush organ are detected by the females' antennae, and if she is in a responsive state, induce mating.

This type of chemical interaction is ubiquitous in the insect world, and indeed forms the basis for many methods of pest control that are environmentally friendly: insect pheromones do not pose ecological hazards of the sort that made DDT so notorious. The chemical specificity of glandular secretions and their corresponding chemoreceptors presumably is designed to ensure that chemical communications are restricted to members of the same species.

Pheromones are also essential to the social organization of many insect species. Pheromones probably play an important, if not essential, role in regulating almost all aspects of behavior in many species of bees and ants. Queen bees secrete substances that regulate feeding and grooming of the queen by the worker bees. In addition to pheromones that promote caregiving, the queen releases sex pheromones that promote attempts at copulation by drones in the colony. Finally, the swarming and attack behavior elicited by disturbances to the hive are mediated by alarm pheromones.

You might think that the behavior of insects is much more automatic than that of mammals, and that, because of mammals' greater sophistication, pheromones might not exert such strict chemical control of reproduction and other aspects of mammalian life. To a certain extent, that may be true. Many would argue that it is especially true in humans, where vision and hearing appear to have gained an ascendancy over the more "primitive" sense of smell. But modes of pheromonal communication certainly occur among mammals, and many experts now think pheromones provide an important source of control in human reproduction and sexual behavior.

Some of the evidence for the existence of pheromones in mammals, especially humans, is controversial, and no reasonable person can discount the obvious importance of visual and higher cognitive processes in human sexual attraction. Much of what has been written about the role of pheromones in human relationships is probably guilty of overinterpreting and overextrapolating what most experts would agree are the scientifically established facts. But there are a number of interesting leads, and there are certainly many puzzles and paradoxes. Regardless of the actual importance of pheromones in human behavior, it should be interesting and fun to try to separate the facts from the folklore.

It does appear that our sense of smell has effects on our behavior and our physiological functions that most of us never imagined. We believe we fall in love with our hearts—or, more likely, our heads. But some would have us believe our noses are an important organ of love, or at least infatuation. More precisely, Cupid's arrows may target a small accessory organ inside the nasal passages called the vomeronasal organ. Where do such ideas come from? Let's take a look and see.

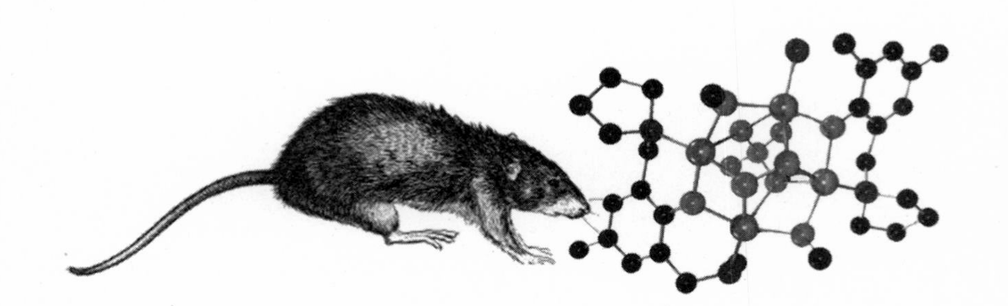

18

Mammalian Pheromones

There is good evidence that pheromones play an important role in the sexual behavior of many mammals. Ultimately, we must consider whether similar mechanisms exist in humans, but as a backdrop for that, let us first consider pheromonal communication in mammals.

Earlier we indicated that pheromones play a role in territorial marking, and there are suggestions that some animals scent trails they frequent by depositing chemical substances through urination, defecation, or rubbing specific glands on foliage or on the ground. But our focus here will be on sexual behavior, because almost all claims of pheromonal effects in humans relate to sexual attraction or reproductive physiology.

Many of the effects in mammals appear analogous to those in insects. For instance, female golden hamsters secrete a vaginal discharge at the time of estrus (the period of female sexual receptivity). This substance excites the male, and stimulates him to attempt copulation. Here we clearly have an effect analogous to the effects of bombykol on male silkworm moths. But male hamsters are not the only ones that are so enslaved by their nostrils. The timing of female ovulation in mice depends upon pheromones present in males' urine. It has been shown, for example, that a group of females housed together have asynchronous estrous cycles until they are exposed to the scent of male urine: then the estrous cycles of all the females become synchronized. Moreover, the onset of puberty can be advanced in females by exposure to the odor of male urine. Thus it is thought that the urine of male mice contains pheromonal substances. Indeed, there is evidence that the nature of the pheromonal substances (or perhaps their relative concentrations) is specific for each individual. In some cases, the concentration of pheromones appears to vary according to the levels of circulating androgens (male sex hormones). Mice can identify the presence of a dominant male by the signature in the odor of his urine.

Because of their popularity as laboratory animals, more is known about the chemical communication systems of rodents than other order of mammal. But that should not lead us to think that rodents are unique in their use of pheromones. Although the exact functional significance in not known, elephants are endowed with glands that are located next to the eyes. These temporal glands are an example of a specific type of skin gland called *apocrine glands.* Apocrine glands are one of three distrinct types of glands found in the skin of all mammals, including humans. In fact, we will discover that humans have a greater density of apocrine glands than any other large primate. Both male and female elephants have temporal glands. Those of the males secrete a fluid when they become sexually aroused. It appears that these secretions likewise excite a female, but only if she is in a state of estrus. And the females also release what may be pheromones in their urine when they are receptive. Astonishingly, these urinary secretions contain substances that are identical to the sexual attractant pheromones used by certain insects! (Rasmussen et al., 1996). Biology can be remarkably economical.

It can also seem fairly odd. Pigs use pheromones, too. But they have their own method of application. When a boar becomes sexually aroused, he secretes copious amounts of a frothy saliva. The saliva contains two steroids that are significant in the reproductive behavior of pigs. Upon an initial (apparently positive) encounter, the boar and sow engage in a great deal of head-to-head contact. If sufficiently encouraged, the male will spray this pheromone-laden saliva all over the sow's face, which induces a rigid, immobile posture in the sow. This posture indicates to the boar that the female is now prepared to mate, and mating then occurs. Interestingly, truffles excrete the same steroids as a male pig, which is apparently why pigs are such good detectors of the subterranean delicacy.

To this point, we seem to have established a pretty clear influence of the nose over sexual behavior. But in certain cases, a sense of smell is more than merely facilitatory; it actually appears essential. For instance, sexually inexperienced mice that are surgically deprived of their sense of smell will *never* engage in sexual behavior. We thus arrive at the idea that there must be a connection, an important and perhaps even essential connection, between the nose and reproductive behavior. Such a connection must exist if, for example, the timing of estrous cycles in females is to be brought under the influence of male odors, as has been observed in mice. There must be some nose-to-gonad connection. In order to see how it works, let's start at the reproductive organs—the gonads—and determine how they are influenced by the brain. Then we can trace the flow of olfactory information into the brain, and when we find where these two pathways meet, we will have established our nose-to-gonad axis.

The reproductive organs contain endocrine glands, the ovaries in females and the testes in males. Like all glands, the gonads secrete hormones—sex hormones. In females, the levels of these hormones vary in a cyclical fashion, and this determines the timing of the estrous cycle—the events that lead to the release of an unfertilized egg from the ovaries. In order to produce offspring, the female must mate when the unfertilized ovum is in exactly the right position along the path from the ovaries to the uterus. So in addition to controlling the release of an egg from the ovaries, sex hormones control the *behavioral* state of sexual receptivity (colloquially known as being in heat).

The question now becomes one of establishing what controls the levels of hormone production in the reproductive glands. The answer is, another gland—the pituitary gland. The pituitary is a small gland located at the base of the brain, essentially right above the roof of the mouth. It is sometimes called the master gland, because it controls all other endocrine glands, including the reproductive glands. The system is somewhat complicated, and really is beyond the scope of what is needed for our current purpose. The essential point is this: the pituitary gland controls all other endocrine glands by releasing its own substances into the bloodstream. These substances are themselves hormones (pituitary hormones). They circulate throughout the body and eventually encounter a target gland (e.g., the thyroid gland, the adrenal glands, the pancreas, the gonads). These pituitary hormones regulate the secretions of the target glands. For example, a pituitary hormone called thyroid-stimulating hormone stimulates the thyroid gland. Another pituitary hormone, called luteinizing hormone, causes the ovaries to produce progesterone in females and the testes to produce testosterone in males. Yet another pituitary hormone, called follicle-stimulating hormone, causes the ovaries to produce estrogen and the testes to produce sperm cells. Since the sex hormones produced in the ovaries and testes in turn control the onset of puberty in both males and females, and as we have seen, the timing of ovulation and the sex drive in both males and females, we see how the pituitary gland is essential to sexual development and sexual reproduction in general.

The central question now is, What controls the pituitary gland? The answer is a small brain structure about the size of a walnut, called the hypothalamus. The hypothalamus is located directly above the pituitary and contains neurons that secrete—you guessed it!—hormones. Many of the hypothalamic hormones are called *releasing hormones* because they regulate the secretion of hormones from the pituitary, which in turn regulates secretions of the other endocrine organs. So the arrangement is one of hierarchical controls. However, the controls are not entirely unidirectional: by sampling the levels of endocrine hormones in its blood supply, the hypothalamus can respond to changes in hormonal levels in the body. In this way, it both controls and is in turn controlled by the endocrine system. The hypothalamus is one part of the brain that differs in males and females—neuroscientists say that certain parts of the hypothalamus are "sexually dimorphic," meaning the anatomy of the structure is different in males and females. (See figure 18.1.)

One hypothalamic hormone that is particularly important in the development and regulation of sexual behavior is gonadotropin-releasing hormone. It controls the secretion of the two gonadotropins, the pituitary hormones that control the secretions of gonads. The pituitary gonadotropins are leutinizing hormone and follicle-stimulating hormone, and we have seen how they control the secretion of sex hormones that are essential to the onset of puberty, the sex drive, ovulation, and successful pregnancies.

One is naturally led to wonder what controls the hypothalamus. (Although this may appear to be an endless regression, fear not! We are

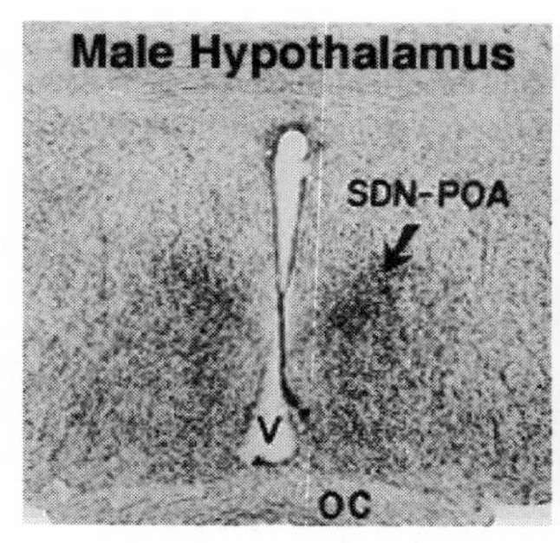

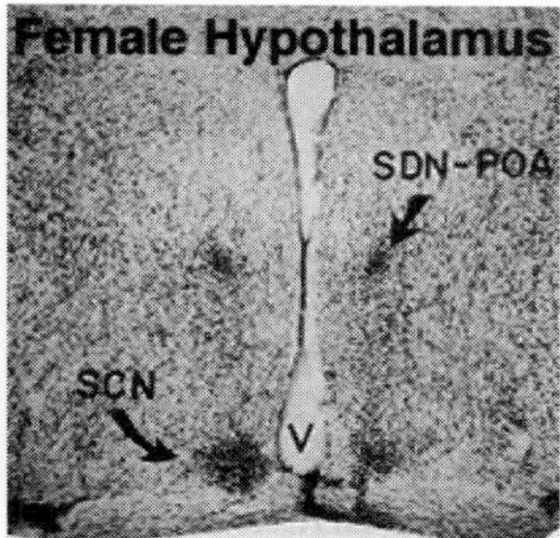

Figure 18.1
Anatomical differences between the hypothalamus in a male and a female rat. SDN-POA stands for sexually dimorphic nucleus of the preoptic area. This brain area is larger in males.

about to close the loop.) We've already encountered one source of hypothalamic control: the endocrine system itself. Since the hypothalamus is equipped with receptors that sample the blood and extracellular fluids, it is informed about virtually any essential body function: from hunger and thirst to body temperature. The hypothalamus also receives extensive inputs from other regions of the brain. It is thus in a position to sample and integrate information from the body and the brain, and to use that information to control many essential organ systems—including, of course, the endocrine system.

Our concern is with the nose, however, so the essential point we must make now is that the hypothalamus does indeed receive inputs from the olfactory system, and this olfactory input completes our nose-to-gonad connection. To appreciate how pheromones might influence endocrine functions, we must elaborate on the details of this nose-to-hypothalamus connection. (See figures 18.2 and 18.3.)

The inside of the nose is more complicated than one might initially think. The nasal passages are lined with a layer of specialized cells. This lining, the nasal epithelium, consists of three distinct kinds of epithelia. Most of the nasal cavity is lined with a respiratory epithelium. It helps cleanse inspired air before it enters the lungs. While that is an important function, it has nothing to do with our sense of smell, so we'll say no more about it. The remaining two types of epithelia are essential to our interests, however. The olfactory epithelium lies near the top of the nasal

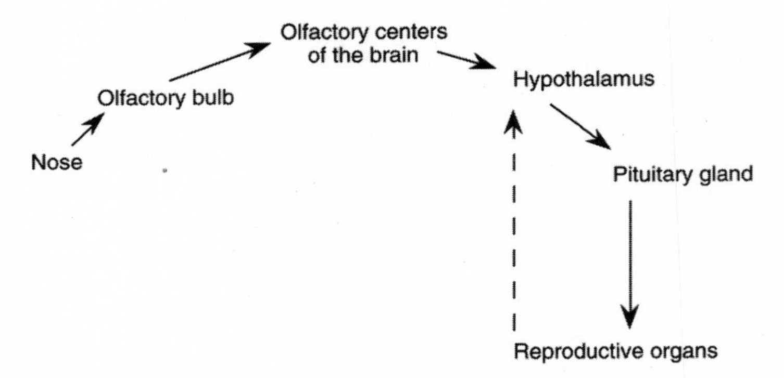

Figure 18.2
Schematic diagram of the relationship between the nose and and reproductive organs. Included is a pathway by which the reproductive organs can influence the hypothalamus (dashed arrow).

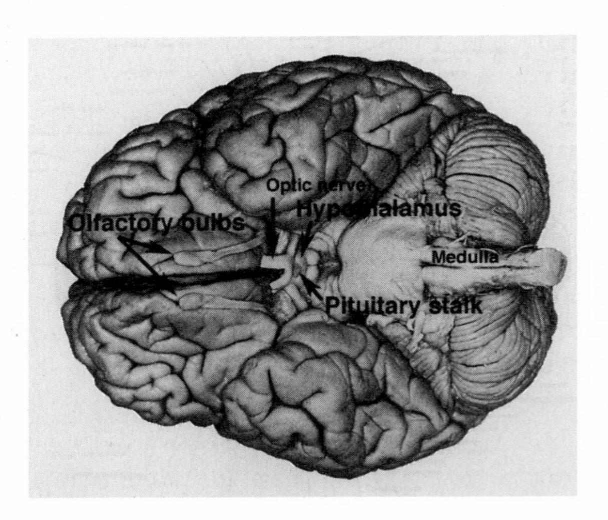

Figure 18.3
A view of the underside of the human brain, illustrating the location of the olfactory bulbs, which innervate the receptors in the nose, the hypothalamus, and the pituitary stalk. The pituitary gland is attached to the stalk, but almost always becomes detached when the brain in removed from the skull.

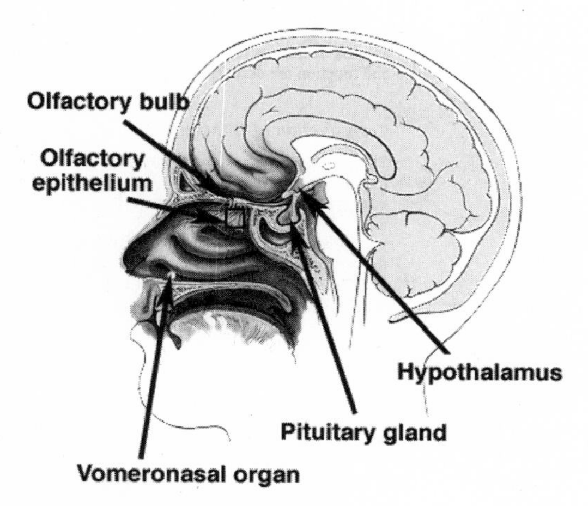

Figure 18.4
Illustration of the locations of the olfactory epithelium and the vomeronasal organ in the human nasal cavity.

cavity (up by the bridge of your nose), and it contains the receptors that permit our "classical" sense of smell—the conscious detection of airborne molecules. These olfactory receptor cells are modified neurons. Their functioning is not well understood. Certainly less is known about olfactory receptors than about visual, auditory, and touch receptors. However, like all sensory receptors, they produce electrical discharges when stimulated. In this case, the stimulation is caused by the introduction of an airborne chemical. Chemicals that produce a conscious olfactory experience are called odorants. Most experts agree that odorants produce their effects only by stimulating the olfactory epithelium—which is to say that the receptors of the olfactory epithelium are *the only* receptors for the sense of smell.

However, virtually all mammals have a third type of nasal epithelium. It is typically located at the base of the nasal cavity, in two small cavities: one in each nostril. These cavities are called the *vomeronasal organ*. The fact that the vomeronasal organ is located at such a distance from the olfactory epithelium suggests that it is a different system, perhaps performing a different function. This suspicion is strengthened by findings that the connections of the vomeronasal organ with the brain are different from those made by the olfactory epithelium. Many experts believe the vomeronasal organ is specialized for the reception of pheromones, and that the actual sense of smell is mediated by the olfactory epithelium. (See figure 18.4.)

Both systems have access to the hypothalamus, although each projects to different hypothalamic subdivisions. The receptor cells themselves do not project directly to the hypothalamus, however. The connections are made via a chain or sequence of neural connections. The receptors send impulses to a structure called the *olfactory bulb*. The olfactory bulb lies on the underside of the front part of the brain. It contains neurons that process the information conveyed by the receptor cells. The olfactory epithelium projects to what is called the *main olfactory bulb*. It is much larger and more conspicuous than the *accessory olfactory bulb*, which receives impulses from the vomeronasal organ. So the olfactory epithelium and the vomeronasal organ each has its individual "receiving center": the main and the accessory olfactory bulbs, respectively.

When the nervous system goes to the trouble of carefully segregating information like this, it is usually for a reason. Different types of information have to be processed differently. Beyond that, different information-specific pathways are often used for different purposes. In the case of olfaction, we suspect that the vomeronasal-accessory olfactory bulb system may convey information that is especially important to those parts of the hypothalamus that are concerned with sexual behavior, whereas the olfactory epithelium-main olfactory bulb system may support the functions associated with detection and recognition of odors in general. In this context, we might expect the differences in brain connections that have been observed for these two systems.

In summary, we have seen that many animal species use chemicals to communicate specific types of information. This information is often essential to the survival of the species, since it relates to defensive and reproductive behaviors. In addition, we have established that if chemical exchanges between individuals are to be used to influence sexual behaviors, there must be a pathway by which the nose can influence the sex organs. The fact that both the olfactory epithelium and the vomeronasal organ provide input to the hypothalamus provides just such a pathway, since the hypothalamus can control the ovaries and testes via its influence over the pituitary gland.

The vomeronasal organ appears essential to pheromonal communication in many "lower" mammals. Rodents have been an especially favored species to study in this context. It is known that males apparently detect chemicals that signal the state of estrus in females and, further, that estrus in females can be "induced" by exposure to male odors. The fact that many of these effects depend on the integrity of the olfactory bulbs confirms the importance of the olfactory system.

These findings lead to the question that has fascinated many scientists, writers, and students of human behavior in general: Might similar effects be at work in human behavior?

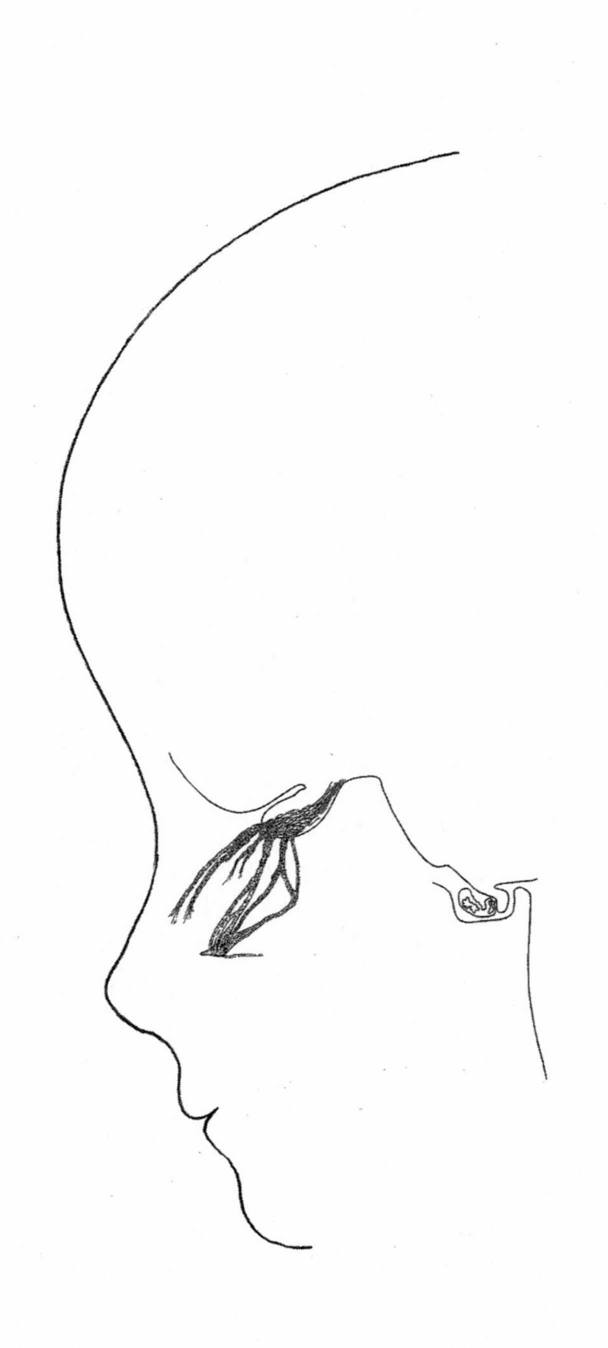

19

Human Pheromones?

When we consider the possible role of odors and the sense of smell in human behavior, and human sexual behavior most specifically, we immediately encounter some difficult problems and intriguing paradoxes. Almost everyone would agree that "social sniffing" is a decidedly "animal" thing to do. Certainly humans do not greet other humans the way dogs greet other dogs. On the other hand, the importance of olfactory cues in human social interactions depends upon cultural variables (Stoddart, 1990). For instance, many are familiar with the traditional greeting between Eskimos: rubbing noses. Few know, however, that this is followed by licking the palms of the hands, and mutual rubbing of hands on one's own and the other person's face. Another example of social sniffing in humans is provided by certain ethnic groups in India, where "kisses" are applied to the nose and cheek rather than the lips. They do not ask for a kiss in their language. Rather, the request to another is, "Please smell me." The European custom of mutual kissing of the cheeks and forehead is interesting when one considers the fact that the forehead, nasal region, and chin are rich in glands whose secretions contain a number of steroid chemicals that, in other species, are known to function as pheromones.

At the same time, any premise that chemicals produced by the human body can have an important influence on sexual attraction and/or reproduction must be reconciled with the obvious fact that, in Western societies at least, most body odors are considered offensive. We wash ourselves with a frequency beyond what is needed for hygienic purposes, and after bathing, apply deodorants and perfumes to further disguise our real body odors. The fact that perfumes contain secretions from the nether regions of musk deer and civet cats—secretions that are almost surely pheromones in these animals—only adds to the paradox. Apparently, we prefer the pheromones of other species to any we might produce ourselves, even though we produce similar (and in some cases, identical) substances.

How might we approach our paradoxical reactions to odors produced by natural body secretions? One tack is to assert that evolution has freed us of the olfactory dominance that is so characteristic of "lower" animal species. Comparative studies of the anatomy of human brains and those of other mammals are often viewed as supporting this view. It is clear that the amount of brain tissue devoted to the sense of smell is *proportionately* much smaller in humans than in any other mammal (with the possible exception of marine mammals, which appear to have especially small "nose brains"). In relation to the their total brain volume, the olfactory bulbs of rodents are enormous. In contrast, the size of our olfactory bulbs is minuscule compared with our total brain size. It seems reasonable to conclude that as our species evolved, vision and hearing became the dominant senses, and olfaction devolved into a vestigial sense; the neglected stepchild of our perceptual systems. It is natural (but egocentric and, more important, erroneous) to view evolution as a linear process that has led to the creation of human beings. We readily fall into the trap of thinking that we represent the pinnacle or current end point of the evolutionary process—it is seductive to think of ourselves chauvinistically as the most evolved species on the planet. The Bible says that we alone were created in God's image, and we readily accept the idea that all other animals can be regarded as "more primitive" than ourselves. As these more primitive life forms have placed a greater emphasis on a sense of smell, it follows that olfaction is a primitive sense. As such, we have little need for a sense of smell. We are "above" all this animalistic sniffing nonsense.

However, if we learn only one lesson from the study of other species, let it be that evolution is not a linear process; it moves in all directions simultaneously. Biosonar and magnetoreception can hardly be regarded as primitive senses that we have discarded. We have simply taken a different evolutionary path, and it is a mistake to try to place divergent evolutionary pathways into a hierarchy in which we have attained preeminence. We are different, which is not the same as being better or more advanced.

Well, if this less egotistical version of our status in the animal kingdom sounds reasonable, then what should we make of the apparent diminishing status of the sense of smell, as reflected in the relatively small olfactory brain of humans? Perhaps we should view the whole matter from another perspective. While our olfactory bulbs are indeed small relative to the rest of our brains, they are enormous in direct comparison with those of rodents. We have not necessarily devoted less brain tissue to the sense of smell. Rather, as our brains underwent an enormous expansion, we devoted relatively more of the additional brain tissue to other neural systems, such as vision and hearing. We also have much more cerebral cortex. And this additional cerebral cortex appears essential for complex thought, mental flexibility, and perhaps conscious awareness. Maybe it is more accurate to say that the olfactory brain has simply not expanded to the same degree as other systems.

Viewed in this way, a significant role of olfactory cues in human reproduction and the selection of mates might not be all that surprising. Yet, whatever the role of olfaction, it seems clear that our reactions to odors are heavily influenced by culture. We don't have the olfactory acuity or sensitivity of many other mammals. But that does not mean olfaction has no role to play. Our reaction to body odors can hardly be regarded as one of indifference. Perhaps the role of olfaction in our lives has been heavily influenced by our cognitive and culturally mediated interpretations of odor. So it becomes uncivilized to sniff one another "like a dog," and it becomes unattractive to convey any of your own body odors to others. Paradoxically, it remains perfectly acceptable, and even attractive, to smell faintly of the pheromones of a musk deer or a civet cat.

In many cultures it is perfectly acceptable to engage in certain forms of "social sniffing." The brains of these people are the same as ours, so it seems misguided to attribute our olfactory paradoxes to a vestigial sense of smell. Western culture's disdain for body odors is likely more a reflection of cultural variables than of the organization of our brains. If this is so, perhaps there are indeed mechanisms of pheromonal communication between individual people; it's just that their role as been obscured and distorted by several thousand years of human culture.

Human behavior is obviously much more complex than that of any other species, and you don't need to read a book to know that men don't detect the presence of an ovulating woman from miles away, and flock to her like so many moths. It's self-evident that any role played by pheromones in human behavior must be much more subtle than the dramatic examples seen in moths, rodents, and pigs. In the face of such complexity and subtlety, an effective approach might begin by asking whether humans have the biological machinery necessary for pheromonal communication. We might ask, for instance, whether people actually possess the glands that produce pheromones in other animals, and whether they possess the sensory apparatus that is typically associated with the detection of pheromones. Of course we have a nose. But we also posses that specialized organ called the vomeronasal organ (VNO). There has been a great deal of controversy as to whether this is actually a functioning organ in the human nose. For a long time it was considered vestigial, but recent evidence indicates that the human VNO functions quite well. Once we have evaluated the findings indicating that humans possess the requisite scent producers and scent detectors, we can turn to consider a variety of phenomena that have been offered as evidence of pheromonal effects in humans. So let's begin with the possibility that the human body produces substances that could contain pheromones.

Animals use a variety of methods for depositing pheromones. We've seen that they include urine-marking, feces, and specialized glands located in the skin. Of these methods, communication via skin glands seems most relevant to humans, so in the search for a human pheromone producer, we focus our attention on the skin. It should be noted, however, that substances suspected of being human pheromones have been identified in urine, semen, vaginal secretions, and saliva.

The skin is the largest organ of the human body, and it is much more complex than we might suspect. Certain "skin facts" may be obvious, but nevertheless should be explicitly stated. For example, the skin is not uniform, but varies over the body surface. Some skin is hairless and some is not. The density and types of sensory receptors differ in different regions of the skin as well. But most important in the context of pheromones is that the distribution of glands varies across the body surface.

There are three types of glands in mammalian skin, and human skin is no exception. Sweat glands are perhaps the most familiar, but skin also contains sebaceous glands and apocrine glands. It is these latter two types that are associated with the production of pheromonal substances in mammals. Both sebaceous glands and apocrine glands are tubular structures that open to the skin surface. More particularly, the ducts of both types open into hair follicles. The sebaceous glands secrete an oily substance called sebum. In most mammals, an important function of the sebum is to coat the fur so that it repels water. There are, however, a wide variety of substances (including small amounts of sex hormones) present in sebum, so the sebaceous glands could in principle form part of the pheromone production system. In an intriguing and extensive account of the role of olfaction in the sexual behavior of humans and other mammals, D. Michael Stoddart (1990) points to the paradox of sebaceous glands in human skin: we are endowed with a high density of subaceous glands despite the obvious fact that we have evolved into a "hairless ape."

The third type of skin gland is the apocrine gland. In many mammals, apocrine glands are known to function as pheromone factories, so they are a natural candidate for sources of human pheromones. They secrete a watery fluid that is laced with a variety of chemical substances. Many of these substances come from a family of steroid compounds called androstenes. Certain androstenes are known to function as pheromones in other mammals, and are considered potential pheromones in humans. They are delivered to the skin surface through small ducts that open into hair follicles. The secretory process of apocrine glands is somewhat unusual in that the cells lining the inside of the gland dissolve into the cavity, so their constituents become a part of the secretions. These cells might thus be the source of human pheromones. In support of this conjecture, we note that apocrine secretions do not appear until puberty, when an individual becomes biologically capable of reproduction.

Perhaps the most intriguing feature of apocrine glands is their distribution in human skin. Unlike sebaceous glands, which are found just about everywhere, apocrine glands are concentrated in a few regions of

Figure 19.1
Distribution of apocrine glands in human skin. The distribution in men and women is similar.

the skin. Our apocrine glands are located on the forehead, around the nose and lips, the ear lobes, the axillary region (armpits), the areolae (area around the nipples), the navel, and the ano-genital region. (See figure 19.1.) In short, areas of the skin that are associated with intimate interpersonal contact, including sexual contact. These are also the regions that, for whatever reason, have retained hair (more so in males than in females). The increased surface area provided by hair can promote the diffusion of chemical substances into the air; thus function of our sparse body hair may be to act like wicks that disperse our pheromones to nearby noses.

The notion that the axillary region is a source of odors that can either attract or repel members of the opposite sex has a long history in human writings and social behavior (Stoddart, 1990). The paradox of human scents and sexual attraction is nowhere more clearly illustrated than in our response to armpits. First, the axillary area is rich in apocrine glands, and is one of the few regions of the body to have retained hair in both men and women. Since it remains moist much of the time, it provides a hospitable environment for bacteria, which act upon the apocrine secretions and produce chemical by-products that have distinctive odors. The axillae thus appear to be *designed* to produce odors. Yet we go to great lengths to minimize those odors, because in most Western societies, they are considered offensive. The practice of shaving axillary hair reduces axillary odor, and one is tempted to conclude that this is one reason why the hair is shaved. We apply perfumes to mask or obscure our body odors. The ultimate paradox is that the perfumes are extracts of the apocrine glands of other animals!

Human skin produces a variety of secretions. Many have odors, but their importance might go far beyond the conscious sense of smell; certain of the substances included in these secretions might act as pheromones. Yet pheromonal communication requires more than chemical production systems, for we have seen that, in order to be effective, pheromones must be able to exert powerful effects on the endocrine system, and on behavior itself. So now we consider how these substances might be detected, and how their detection might influence *our* behavior and *our* endocrine systems.

The human nose is the Rodney Dangerfield of sensory systems: it just doesn't get much respect. It's true that our nose is not nearly as sensitive as, for example, that of a dog. The conventional wisdom is that while the nose reigns supreme among modalities in rodents, canines, and other mammals, it has been eclipsed by vision and hearing in humans. Thus, perhaps the nose is destined to the same fate as the appendix or the tonsils. Perhaps, in time, it will become a vestigial apparatus, useful only for supporting eyeglasses, or the display of jewelry.

An alternate version of the natural history of the human nose could argue that it has retained a place of uniqueness and distinction in human perception—but one that has been obscured by the power of our other senses, and by the evolution of human culture. Most would agree that the sense of smell does not convey the rich information about our surroundings that vision does. But in the minds of many people, olfactory sensations have an especially important impact on our emotions and our memory. I have, for instance, the following impression. See if any of this sounds at all familiar.

When I was a young teenager, we had dances in the school gym. This was a time when Diana Ross and the Supremes reigned. All the girls wanted to look like the Supremes, and tended to wear their hair the same way—in a sort of beehive-shaped sculpture. In order to fix their hair in this way, the girls had to tease it into a big ball, and then smooth it over into the round shape. This required a lot of hair spray, and the girls used it in great quantities. And, just as the hairdos were all the same, so all the girls at my school used the same hair spray. I don't know what kind it was, but when we danced cheek-to-cheek, I always got a good whiff of this hair spray. Sometimes it even got in my eyes. That burned, but I didn't care, because I was dancing with a great girl with a very stylish hairdo. The hair spray had a distinctive odor. It was unforgettable. I haven't smelled that hair spray in many years, but I am confident that I would immediately recognize it if I were to encounter it again.

Many odors seem to have that quality. You can't really describe them, you can't really imagine them. We have an expression of mental imagery—we call it the mind's eye. I have never heard anyone refer to the mind's nose. But there is one thing about odors—you remember them. Even if you hadn't smelled a particular odor for many years, you would immediately remember it upon first whiff. Many experts point out that smells can immediately access memories that may have lain dormant and undisturbed for thirty or more years, kind of like hearing an old song.

The other thing about odors is that they can evoke powerful emotional responses. As Stoddart (1990) points out, this is reflected in our common expressions. Why is it, for example, that vision is so often used as a metaphor for understanding ("Do you *see* what I mean?"), but so often odors are associated with (negative) emotions ("What a *rotten* thing to do," "That really *stinks*," "I *smell* a rat")?

The close association between smells, memory, and emotions might have a neuroanatomical explanation. The olfactory bulb projects into areas of the brain that we know contains neural circuits involved in memory and emotional expressions. Indeed, the classical neuroanatomical term for the structures that contain these circuits is the *rhinencephalon* (nose brain). The rhinencephalon includes the olfactory bulbs, the amygdala, and the hippocampus. We've already seen that portions of the amygdala are concerned with the processing of olfactory information, and that the amygdala provides a pathway to the hypothalamus, which in turn controls the endocrine systems via the pituitary. However, the amygdala is also associated with the production of emotional states and the formation of memories. In fact, it is an integral part of a network of brain structures involved in the production of both emotions and memory. Experiments in animals have shown that electrical stimulation of portions of the amygdala can produce dramatic displays of aggressive behavior. Conversely, bilateral lesions of the amygdala produce a taming effect. These latter findings led to the psychosurgical practice of performing bilateral amygdalectomies in the criminally insane during the 1940s and 1950s. The procedure did indeed produce the desired taming effect, but has since been replaced by the use of pyschoactive drugs. Bilateral damage to the amygdala and surrounding structures also has the curious effect of "releasing" sexual urges in certain cases. This hypersexuality syndrome has led to the suggestion that the amygdala is concerned with the inhibition of sexual activity. In addition, bilateral damage to the amygdala and the adjacent hippocampal formation produces a well-known amnesiac effect.

So the amygdala is involved in an especially diverse set of biological functions when you consider its small size (about the size of your thumbnail in a human). These functions include (but are not necessarily limited to) a modulating influence over many emotional states, the formation of memories, and learning. Although other sensory modalities eventually gain access to the amygdala, the olfactory sense is the only one that enjoys direct access. Perhaps these arrangements reflect the central importance of the sense of smell in our ancient ancestors.

Pheromonal communication requires the detection of chemical compounds produced by another individual. While this leads us to focus on the sense of smell, it turns out that pheromone detection appears to be performed by a system that is different from the one we typically associate with the "ordinary" sense of smell. Olfaction begins when an airborne molecule comes in contact with receptor cells that lie in the depths of the nasal cavity. The molecules interact with the receptor cells in such a way as to cause ions to flow across the receptor cell's membrane. These ions have an electrical charge, so their movements constitute electrical currents. These currents stimulate a cascade of electrical activity that enters the brain and ultimately produces an olfactory sensation.

In a real sense however, many animals have two noses—the olfactory epithelium and the vomeronasal organ. Each has a distinctly different population of olfactory receptors, and each is located in a different region of the nasal cavity. The "typical" olfactory receptors are located in the upper region of the nasal cavity, very close to the base of the brain. In contrast, the vomeronasal organ is located at the base of the nasal cavity, on either side of the nasal septum (the structure that divides the left and right nostrils). The vomeronasal organ is very small—usually less than a millimeter across. But its importance is far greater than its small size might suggest, for this third type of nasal epithelium is responsible for the reception of pheromones.

The vomeronasal organ has its own receptor cells that are anatomically and functionally different from those in the olfactory epithelium. The activities of vomeronasal receptors are conveyed to the brain over a different nerve than the more familiar activities of the olfactory receptors, and the vomeronasal and olfactory nerves project to different regions of the olfactory bulb. Thus, the idea of two noses is not much of an exaggeration—there are two distinct types of receptors, and each has its own neural processing system. The olfactory receptors contribute to what is termed the main olfactory system, whereas the vomeronasal receptors provide the input to the accessory olfactory system. In rodents and other mammals, the vomeronasal receptors and the accessory olfactory system appear to be essential for pheromonal communication and for the expression of normal reproductive behavior (Halpern, 1987).

The vomeronasal organ was first identified in humans by a Dutch anatomist named Frederik Ruysch in 1703. Almost 300 years later, its role in humans remains mysterious. Indeed, until very recently, the VNO was thought to be completely vestigial in humans. Although anatomists could routinely see the organ in fetuses, its presence was only occasionally confirmed in adults until the 1980s, when investigators seem to have learned how to unambiguously identify the VNO in adults. The human VNO appears as a small opening in the nasal septum. This small pore is found toward the front and base of the nasal septum and provides the access to a small invagination that is lined with cells that look like sensory receptors. Anatomists have been unable to verify unambiguously that these cells receive a neural innervation, but nerve fibers do come quite close to the suspected vomeronasal receptors.

A recent series of electrophysiological studies suggests that the human VNO does indeed produce electrical responses to certain chemical compounds. The work has been done at the University of Utah School of

Medicine by Luis Monti-Bloch and his colleagues, some of whom work for the Pherin Corporation of Menlo Park, California. Dr. Monti-Bloch has developed an electrode that permits precise delivery of small amounts of airborne chemicals to either the vomeronasal pit or the main olfactory epithelium. The device consists of a thin silver wire threaded through a small teflon tube. This tube is inserted within a slightly larger tube, so the electrode lies within two concentric tubes. The inner tube is connected to containers that hold different chemical compounds mixed with air. The stimuli consist of small amounts of these compounds that are extruded through the inner tube. Sensory responses are recorded via the silver wire. A weak vacuum is applied via the outer tube, so the chemicals that do not interact with receptors are immediately removed. The entire device is inserted up a volunteer's nose. The tip can be placed near the olfactory epithelium, the vomeronasal pit, or the respiratory epithelium, so responses to the compounds can be compared.

Very interesting results with this device have been reported by Monti-Bloch and colleagues. First, as expected, no electrical responses to any compounds are observed when the stimuli are delivered to the respiratory epithelium. Neither do the subjects report any conscious olfactory sensations. These observations are important in that they demonstrate that the vacuum system performs the job it was intended to perform, which is to prevent the stimuli from diffusing throughout the nasal cavity. Second, electrical responses are observed when "standard" odorants, such as lemon and citronella, are applied to the olfactory epithelium. This proves that the electrode is capable of recording responses from olfactory receptors. Along with this activity, people report detecting the odor. So far, all observations are as expected and indicate that the system works well. Receptor responses are observed by using the electrode, and the stimuli appear to be confined to a small area around the electrode, as intended.

The most interesting results involve recordings of near the VNO. No detectable responses are obtained using the standard odorants. In addition, application of standard odorants to the vomeronasal opening does not produce any olfactory sensations. Responses are obtained when the investigators apply putative pheromones to the vomeronasal pit. Now we must mention one small caveat here. The investigators have not disclosed the chemical structure of the compounds that produce these responses in vomeronasal receptors. The work was supported by the Pherin Corporation, and the investigators state they will disclose the exact formulas for the compounds when their application for a patent is complete. Nonetheless, they do state that the compound is a steroid molecule, and that it has been extracted from human skin. They refer to the substances that induce electrical responses in the human VNO as "vomeropherins," and they regard them as putative (suspected) human pheromones.

Application of vomeropherins to the olfactory mucosa produces no detectable effect: the vomeronasal receptors appear to be the only ones that respond to these potential human pheromones. In addition, humans do not report any conscious olfactory sensations when vomeropherins are applied to their vomeronasal organs. Apparently, the sense of smell as we typically think of it is the exclusive province of the olfactory epithelium, but certain compounds produced by human skin can stimulate vomeronasal receptors. We may remain unaware of these responses.

According to Monti-Bloch and colleagues, out of mind does not mean out of brain however. In additional to recording vomeronasal responses, these workers measured heart rate and skin conductance. Vomeronasal receptor responses were accompanied by changes in heart rate and skin conductance, despite the fact that subjects reported they did not "smell" anything. In a subsequent study, a synthetic vomeropherin not only produced vomeronasal receptor responses along with changes in heart rate and skin conductance, but also influenced the release of gonadotropic hormones in males (but not females). Thus, stimulation of the vomeronasal receptors can influence the hypothalamus and the reproductive organs, and the same compound can produce different effects in men and women.

Additional work is clearly needed. Although the experiments of Monti-Bloch and colleagues appear technically sound and the results appear unambiguous, confirmation of these effects from a different laboratory is needed. However, the available data indicate that human pheromones might be produced by the skin and detected by a functioning VNO. The potential for pheromonal communication is there. We turn now to consider some phenomena that suggest the operation of pheromones in humans.

There are several lines of evidence that are at least consistent with the idea that olfactory cues play an important role in human reproduction. There is even evidence that olfactory cues may play a role in sexual attraction. All the evidence is circumstantial, and each point is subject to alternative explanations. Yet, when taken in the aggregate, the findings make an intriguing story. Caution is certainly in order, and any firm conclusions must await more definitive findings.

To this point, we have outlined the evidence that humans have the machinery that might be expected both to produce pheromones and to detect their presence in other individuals. Now we turn to the third leg of our story: evidence for actual pheromonal communication.

In 1971 M. K. McClintock reported that social factors influence the timing of menstrual cycles in groups of young women. McClintock asked a group of incoming female college freshmen to fill out a questionnaire soon after their arrival on campus. The women were asked about the time of their last menstrual cycle, and to provide additional information concerning the regularity of their monthly cycles. These women were all assigned to the same dormitory and, at first, had asynchronous menses. Within four months however, the menses of many of the women had become synchronized, so that their ovulatory cycles occurred at about the same time of the month. Many women may have experienced similar entrainments of their cycles with those of friends or coworkers. We know that this induced synchrony of ovulation must be mediated through hypothalamic control over the pituitary gland, and it evidently depends upon some form of social or personal contact. Olfactory cues seemed a promising candidate. Indeed, a recent report by K. Stern and M. McClintock (1998) demonstrates that exposure to odorants extracted from axillae of donor women produces menstrual synchrony. The synchronizing effect occurred even though the compounds produced no noticeable odor. A similar study by McClintock indicated that exposure to males can significantly reduce the temporal variability between menstrual cycles. The exposure was entirely of a social nature; intimate contact was not necessary for menstrual synchrony.

In the absence of intimate contact, it seems natural to suggest that olfactory signals may play a role. But like much of this work, there are conflicting results. For example, menstrual synchronization appears to depend on the social relationships between women as much as their physical proximity: synchronization occurs much more readily between best friends than between neighbors or roommates who are not close friends. This would not be expected if the effect were purely pheromonal in nature, and could be used to argue against a pheromonal mode of action. The data are inconsistent, however. Consider an experiment conducted by Russell, Switz, and Thompson (1980). The purpose of the experiment was to determine whether olfactory cues alone can influence the timing of human menstrual cycles. The experimenters carefully selected a "donor woman," who had earlier noticed that her roommates tended to synchronize with her. The experimenters then placed cotton pads in the axillae of the donor. She had stopped shaving her armpits, and had ceased applying deodorants. After she wore these cotton pads for twenty-four hours, a small amount of alcohol was added to the pads, which were then frozen. Sixteen female volunteers served as experimental subjects. The cotton pads were dabbed on the upper lips of the subjects three times a week for four months. Eight of the women received pads that had the axillary secretions of the donor, and eight received pads that contained only the alcohol. The subjects did not know whether their pads contained secretions from the donor woman or not. All subjects kept a diary of their menstrual cycles. The women exposed to the donor's secretions became entrained with her, whereas the timing of the cycles in the control group was random.

Similar effects have been reported when extracts of male axillae were applied to the upper lips of female volunteers: variability between the women's' cycles was reduced, and the cycles became more regular. Further evidence in favor of the view that axillary secretions of men influence the reproductive physiology of women comes from studies in which ovulation is the focus, rather than the menses per se. Women who are celibate do not ovulate during every menstrual cycle—in fact, they ovulate only about half the time. Women who have contact with men, however, ovulate 90% of the time, and their cycles become more regular. Veith and coworkers report that these effects are unrelated to the frequency of sexual intercourse, suggesting that ovulation is to some extent under of the control of an unidentified male pheromone.

In his fascinating and very scholarly book *The Scented Ape,* Stoddart (1990) relates much of the folklore associating the nose with sexual behavior and reproduction. Practitioners of the art (or pseudoscience, depending on your perspective) of physiognomy often assumed a relationship between the size of the nose and the size of the penis. Perhaps it is this assumed relationship that explains the ancient practice of punishing adulterers by amputating their noses! It was once believed that one could verify female virginity through careful manipulation and inspection of the nasal cartilage.

A clinical condition known as Kallman's syndrome provides a particularly dramatic example of the link between the nose and the reproductive organs. Kallman's syndrome is characterized by a lack of normal sexual development. In males, the testes remain small and the axillae remain hairless. In women the ovaries and breasts do not develop fully. In short, these patients never complete the process of puberty. Not surprisingly, the condition is often attributed to dysfunctions of the endocrine system. These patients also invariably suffer from anosmia—they can't smell. This association between underdeveloped genitalia and the lack of a sense of smell was first noted by a Spanish doctor named Aurelio Maestre de San Juan y Muñoz in 1856. One hundred and thirty years later, magnetic resonance images of the brains of patients with Kallman's syndrome show they lack olfactory bulbs (Klingmuller et al., 1987).

The evidence of a nose-to-gonads relationship naturally raises the possibility that olfactory cues may play a role in sexual attraction as well as sexual development. Again, the strongest evidence comes from studies of rodents. For example, Fillion and Blass (1986) found that in male rats, odors associated with the mother during nursing heavily influenced sexual attraction in adulthood. These workers gave some nursing mother rats the odor of lemon while others retained their natural odors. Males that as pups had nursed the lemon-scented mothers preferred to mate with lemon-scented females, whereas males that had nursed naturally scented mothers preferred naturally scented females over lemon-scented females. It appears that in many species, exposure to particular odors early in life produces lifelong preferences for those odors (Stoddart, 1990).

Are similar preferences at work in humans? This has been somewhat more difficult to determine. There have been, for example, a number of attempts to determine whether odor cues influence our perceptions of other people. When considering the issue of scents and interpersonal attraction, it is important to keep several points in mind. First, many different substances are present in the secretions of sebaceous glands and apocrine glands: which (if any) of these substances are pheromones has not yet been determined. Second, pheromones may or may not have a detectable odor. Many of the steroid compounds that can be identified in axillary or other secretions do have a detectable odor, but it is generally rated as unpleasant. One the other hand, some compounds may stimulate receptors within the VNO but not produce a conscious olfactory sensation (perhaps because they may stimulate vomeronasal receptors without producing activity in the olfactory epithelium). So we don't really know which compounds we should test to see whether they influence interpersonal perceptions, and we don't know whether or not any effective substance would produce detectable odors. Finally, any experimental test of pheromonal influences on human behavior is at very best a weak approximation of complicated interactions between a multitude of variables that naturally influence sexual attraction in humans. It is always possible that certain essential variables are omitted in translating the situation from real life to a controlled laboratory experiment.

A variety of experimental attempts to study the influence of body odors on interpersonal judgments have been made. Taken together, the results suggest that some effects may be present, but their exact nature remains unclear. The paradigms used to search for effects of scents on attraction are necessarily unusual. Obviously, the subjects must sniff something, and then rate the odor in some way. In a typical experiment, we attempt to isolate the one variable of interest: odor, in this case. It would be a bad experimental design to have subjects actually sniff another person and rate his or her attractiveness, because the ratings could easily be influenced by other cues, like vision. So most experiments take extracts from a donor, and have a volunteer subject rate the pleasantness of the odor. The donor is never seen by the rater. Many experiments have the donors wear a cotton shirt for a day or two, and the odors of the shirts are then rated. In several experiments, the odors of vaginal secretions have been rated. In these cases, secretions are typically placed on a cotton swab or some "neutral" material that the subjects can sniff. Most experiments have found that subjects rate almost any body odor as essentially unpleasant. The stronger the odor, the more "unpleasant" the rating. So far, no one has found a natural body odor that really pleases other people. There is, however, evidence that odors from different people are discriminated.

For example, babies can reliably discriminate the odor of their mother's breast from the odor of some other women's breast. Conversely, many mothers (and fathers) can reliably identify clothes worn by their own baby from those worn by other babies. Of course, these effects need not be examples of pheromonal communication: these discriminations might be based on more traditional olfactory cues.

In a frequently cited paper, Cowley and colleagues (1977) studied the effects of two odorants: the steroid compound androstenol, and a mixture of what are termed aliphatic acids. Androstenol is found in axillary secretions, and is produced in the testes of human males. It is also the pheromone that produces the immobile posture assumed by female pigs prior to mating. Aliphatic acids are found in the vaginal secretions of humans and rhesus monkeys. These acidic compounds promote sexual arousal in male rhesus monkeys. The subjects in the Cowley et al. experiment were asked to rate six individuals, three males and three females. Subjects were told that the six individuals were applicants for a job in a university student union. Subjects did not meet the "applicants" in person, but were asked to provide ratings based on photographs and text describing some personal details of each applicant. In what might seem a contrived procedure, the raters were required to wear surgical masks. They were told that the masks were intended to prevent the other raters in the same room from seeing their facial expressions. It might be supposed that this would raise suspicions in some subjects regarding the real goal of the study. Some of the surgical masks were treated with odorants, and some were not. The goal of the experiment was to determine if the odors had any effect on the ratings.

Effects were found, but they were modest in magnitude, and the pattern of results was quite complicated. Males exposed to the aliphatic acid mixture gave *less* favorable ratings of one of the female candidates, but they tended to rate this candidate less favorably in general. In no case did the aliphatic acid mixture act to increase favorable ratings of any of the three female candidates. Indeed, when compared with male subjects who wore unscented masks, these extracts of vaginal secretions lowered average ratings of all three female candidates—not the outcome one might expect if the acid mixture was a sexual attractant. The acid mixture did produce a modest increase in favorable ratings of female candidates by female subjects, but again, this might not be expected if the aliphatic acids somehow act to enhance the attraction of males to females. There was also little to suggest that the androstenol enhanced the favorable ratings of male candidates, by either male of female raters. In the last analysis, it seems that the individual differences in the candidates had the greatest influence over positive ratings. Whatever effects were attributable to the odorants differed for different candidates, and depended on the gender of the rater. While it might be expected that the different odorants influence males and females differently, it is difficult to see how the pattern of results actually obtained is consistent with either compound's acting like a universal sex attractant.

Filsinger and colleagues (1984) also found that androstenone can influence the perception of other people. In addition, they found that the odorant could influence people's ratings of their own moods. Androstenone had different effects on men and women, but, as in the experiment of Cowley and colleagues, the effects were complicated and did not conform to a pattern that resembles enhanced sexual attraction. In fact, the odorant reduced self-ratings of sexiness in the female subjects.

Benton and Wastell (1986) felt that these attempts to determine the effects of androstenol on human sexual behavior lacked what psychologists call "face validity." That is, they examined androstenol in situations where sexual arousal might be considered inappropriate, or at least very unlikely. It seems a fair objection. After all, most people would probably try to inhibit or dismiss any feelings of sexual arousal that emerged as they examined dossiers of job applicants. So Benton and Wastell took a more direct approach. They had 100 female subjects read a passage from an adult magazine (*Penthouse*). The passage explicitly described a sexual encounter between a man and a woman on a train. A control passage described the life of a journalist. The "surgical mask technique" was used. Half the masks were treated with androstenone, and half were not. Half of the subjects were randomly assigned to read the sexually arousing article, and half read the story about the journalist. Subjects were told that the surgical masks were part of an experiment on breathing rates, and the authors report that no subject expressed suspicions about the real purpose of the masks. Half of the women assigned to read the *Penthouse* article wore masks treated with androstenol mixed with ethanol, and half wore masks treated with ethanol alone. The women assigned to read about the journalist were similarly divided into an androstenol group and a control group. After reading her assigned passage, each subject completed a rating scale designed to measure her mood. The scale included ratings on a variety of mood dimensions, including Dominant-Submissive, Happy-Sad, Relaxed-Tense, Not Sexually Aroused-Sexually Aroused. Self-reported feelings of sexual arousal were higher in those subjects who had read the *Penthouse* article than those who had read the control article. However, none of the ratings were influenced by androstenol. Benton and Wastell discuss the possibility that uncontrolled variables (like the stage of each women's menstrual cycle) may have obscured a small effect, but they acknowledge that sexual arousal was achieved by the reading material, and that androstenone had no identifiable effect on that response. There is little evidence to indicate that the human response to these substances in any way approximates that seen in other mammalian species. So far as I am aware, we have no scientifically accepted evidence of a universal sexual attractant pheromone that influences the behavior of either men or women.

These findings might disappoint those who find the idea of an olfactory sense of attraction intriguing. Like those who would have you believe that a single additive—a pheromone derived from human skin—could end loneliness, alienation, and bachelorhood, simply by adding a few drops of the incredible extract to your perfume. The ultimate, natural love potion. The enigma of apocrine glands, with their interesting distribution over the skin surface and their secretions that include known pheromones in other species remains with us, however. And why is it that these glands don't begin to secrete anything until puberty? Just as elusive is the importance of an apparently functional VNO in human adults. Why do we need all this equipment if it doesn't permit some subtle form of communication between individuals?

Maybe we have been looking at this from the wrong way around. Perhaps there is a chemically mediated sense of attraction, but the notion of a universal sexual attractant—this whole idea that one male and/or one female odor fits all—is simply not applicable to human reproduction. After all, our sexual lifestyles appear to have some pretty substantial differences from those of most other species of mammals, and the language of chemical communication may have more than just one word in it. Perhaps the biological strategies underlying human reproduction are much more complicated than that.

Consider, for example, that most animals reproduce seasonally, whereas humans can reproduce at any time of the year. Maybe the fact that human females can be sexually active throughout the year reduces the need to secrete potent chemical signals that advertise their "readiness" to mate. A female moth that has become sexually receptive may have only a brief opportunity to mate in her whole life. If the males and females are widely dispersed, the first goal must be to bring them together. Under such circumstances, powerful sexual attractants have an obvious advantage. Selection pressures will favor both the production of copious amounts of the attractant in the female and an exquisitely sensitive detection system in the male. In this way, the chemical communication can occur over great distances. Similarly, many mammals are solitary by nature, and encounters between males and females may occur only during the mating season. Again, the value of chemical messengers that inform a male that a receptive female is in the vicinity seems apparent. We might interpret the ability of male odors to induce sexual receptivity in females along similar lines.

But when we consider situations in which males and females are routinely found together, and they mate just about anytime, the situation may become more subtle. Sexual lifestyles differ in different animal societies. In some cases, access to sexually receptive females is the sole province of a dominant male. The dominant male is likely to be the biggest, strongest, and most aggressive male in the colony. In that sense, it seems perfectly reasonable to form a society in which many females will be fertilized by the same male—the dominant male is likely to be the most fit. All the females can mate with the dominant male, and be reasonably assured that they are receiving "good" genes—genes that will maximize the likelihood of success of their offspring. The females in this case do not need to compete with other females—the dominant male will happily mate with all of them! So they do not necessarily need to attract the male, but they may need to advertise the fact that they are ovulating. Changes in external appearance and certain chemical signatures might be used for that purpose.

In certain primate societies, females mate with many males quite routinely. If each female can receive the sperm from many males, selection of the best sperm obviously doesn't come from behavioral rules concerning who couples with whom. It's a little late for that choice. Choices among males might still be possible, however. There is evidence, for example, that there are many different types of sperm cells, and some don't even carry genetic material. It seems that not all these sperm are designed to fertilize eggs. Some, for instance, are "killer sperms." Their job is to find as many sperm from other males as possible, and kill them all. As all this sperm warfare takes place inside the female's body, postcoital selection is taking place. Rather than using behavioral means to ensure that only the best genes are used to create progeny, in some species the sperm cells from many males fight it out among themselves, and only the most fit sperm succeed in fertilizing the egg.

It is obviously difficult to determine exactly where humans fit into this spectrum of reproductive strategies, although almost every known human culture has some form of formalized pair bonding (marriage). The origins of the institution of marriage probably represent a response to the prolonged period of dependence in human babies. This requires a steadfast commitment of at least one adult, usually the mother. Such an extended commitment may have led to division of labor such that some adults take care of the babies, and some take care of food gathering, house making, and the like. In some animal groups, many females collectively care for all the young. If that will work, then the males need not hang around very long. But that is likely to work only if all the babies are about the same age—caring for babies of all ages over a period of ten or so years is a lot to endure for any adult who is not actually related to the kid. Enter Dad? Pair bonding? Marriage?

In any case, lifelong pair bonding is a goal (if not an actual practice) of many human cultures. So we're not going to go around having coitus whenever the urge arises, and we're going to have to pick a mate—one we're likely to be with for some time. One who will provide half of our children's genes. Good ones, we hope.

Viewed from the perspective of the continued evolution of our species, a good mate increases the probability that his or her genes are preserved in living offspring. A good mate maximizes the chances of successful reproduction. A good mate has a good immune system, or at least one that complements our own.

What does the immune system have to do with all this? Twenty years ago, that's what almost anyone would have asked. But not today. Today there is evidence that odor cues can convey important information concerning details of that portion of the genetic code that controls the biochemistry of the immune system. In mice, we know the relevant portion of the DNA is a group of genes called the major histocompatibility complex (MHC). The MHC controls the manufacturing of proteins that are embedded in the cell membranes of the immune system, cells that fight infections. The same cells produce the rejection response in organ transplants (thus the term *histocompatibility complex*).

The MHC turns out to be a place where genetic mutations are pretty frequent (relative to other gene complexes). As a result, there are many different versions of the complex in different individuals of the same species. Each version confers on its owner a set of responses that the immune system can make. Some immune systems are especially resistant to one set of diseases, and others may be particularly effective against other diseases. The thing is, the baby's MHC will be a hybrid that will have the best features of MHC from each parent. It seems that a good reproductive strategy would be to form a pair bond with another individual whose MHC complements your own. That way, your kids are more likely to get a better immune system than you have. And that will help keep us one step ahead in this microbe-infested world. One step ahead of the germs that make us sick, and frequently kill us. So how might one individual know whether a potential mate's MHC is compatible with his or her own? Maybe he or she can smell it.

In the late 1980s, a group of researchers working with an inbred strain of mice found that mice that were genetically identical except for the region of the MHC complex could discriminate one another by means of olfactory cues alone (Beauchamp, Yamazaki, and Boyse, 1985). Thirsty mice were taught to select one of two arms of a maze on the basis of small concentrations of odorants introduced into each arm. Once they succeeded in selecting, say, lemon scent or cinnamon, they were given harder tests. For example, odors from two different mice. In order to get a water reward, the trained mice had to select the arm that had the "correct" individual odor. Amazingly, while the odors from two genetically identical mice could never be discriminated by another, mice that were identical in all respects except for their MHC complexes could always be distinguished—by their odor alone. It's as if the mice can identify another's MHC complex by odor cues alone.

Additional studies showed that these olfactory cues play an important role in the reproductive behavior of the mice. As part of the inbreeding regimen needed to generate varieties of inbred strains, a male, a genetically identical female, and a female whose genes differed only at the MHC site were placed in the same breeding cage. In these situations, it was found that the mice with dissimilar MHC complexes were more likely to mate and successfully reproduce than mice with exactly the same genotype. This preference for mates whose MHCs differ makes sense because it improves the immune system of the next generation. And mice can make the distinction by using olfactory cues alone.

Then the other shoe fell. It was discovered that this ability to discriminate odor differences was not restricted to the mice: humans could discriminate the odor of mice that were genetically identical except for their MHC. The finding suggested the possibility that humans can discriminate the odors of other humans based on MHC genotypes.

The question was addressed by using a relatively standard procedure. Forty-nine females and forty-four males were typed for antigens determined by their MHC. The subjects were students at the University of Bern in Switzerland. The males were asked to wear an untreated cotton T-shirt for two nights. They were required to use perfume-free soaps, and perfume-free detergents to wash their clothes. They were asked to avoid eating odorproducing foods (a list of such foods was provided), smoking, drinking alcohol, and sexual activity. At the end of the two-day period, each T-shirt was returned to the experimenters.

Each female subject was then asked to smell six of the T-shirts. Three of the shirts were selected from men whose MHC was different from that of the female being tested, and three were similar to it. The women had no way of identifying the donor of any given T-shirt. Each was asked to rate the odor of the shirt for intensity and for sexiness. The women were asked to use a nasal spray that promotes regeneration of the olfactory receptors.

Remarkably, the results showed that women's ratings of the sexiness of a given shirt depended on the similarity of the rater and the donor. The less similar the two MHCs, the greater the rated degree of sexiness. But this pattern was reversed in women who were taking birth control pills. In the latter, similarity in MHC produced higher ratings of odor pleasantness than MHC dissimilarity (Wedekind, Seebeck, Bettens, and Paepcke, 1995). The authors point to the implied paradox of birth control pills: they may interfere with optimal selection of a mate! They also describe findings that couples who have trouble conceiving, and have failed after two or more attempts at artificial insemination, tend to have a greater MHC similarity score than couples who succeeded on the first attempt at artificial insemination. Couples who have similar MHC scores also may have an increased incidence of spontaneous abortions, and the babies they do have, tend to have lower birth weights. The authors conclude that "MHC-correlated abortions . . . could result from strategic 'decisions' of the woman's physiology about her investment in her baby . . . a 'decision' that is most probably no more conscious than, for example, the 'decision' to reject an allograft." So, despite what the advertisements might tell you, masking your odor with perfumes, deodorants, and scented soaps might impair your ability to find an optimal mate!

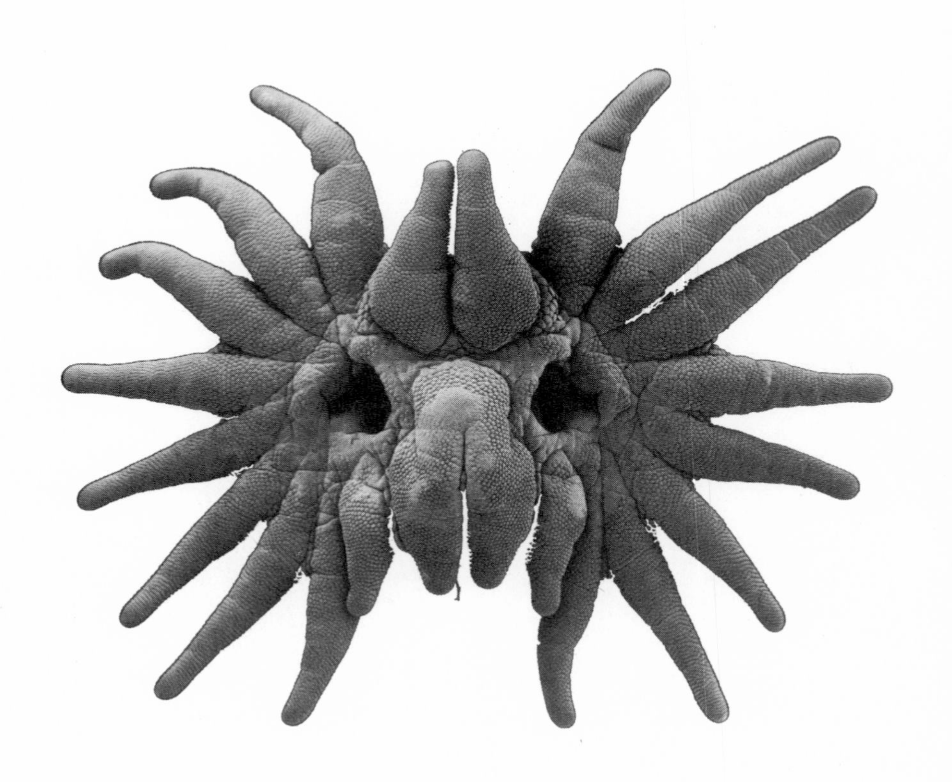

20

Epilogue

What a long, strange trip it's been
Jerry Garcia, "Truckin'"

Investigations into the nature of the senses have been ongoing for hundreds or even thousands of years, depending on your definition of the beginning. Regardless of definition, we certainly have been engaged in the study of the senses for a very long time. We are finally getting good at it.

By almost any criterion, our journey has had its share of strange episodes. Bats have been smeared with Vaseline to determine whether their nighttime navigation depends on a sense of touch. Modern medical imaging technology has been used to determine the anatomical source of the sonar signals of dolphins. Remarkably, it was discovered that these ultrasonic signals are emitted from their foreheads. We've raised birds under artificial skies created by planetariums to demonstrate that they can navigate using the stars. We have demonstrated that women prefer the odor of males whose immune system complements their own—by asking them to sniff T-shirts worn by strangers! Who would have guessed the trail of discovery would lead us to perform such seemingly bizarre experiments? The facts are often stranger than fiction.

Our topics have indeed been wide-ranging, but several general themes have emerged. First, these sensory abilities, no matter how exotic, are all based on reception of real, measurable physical energy. In each case, we found that this energy must first be detected by specialized receptor cells.

These cells act as transducers, converting the physical energy into minute electrical currents produced by the movements of ions across cell membranes. In turn, these currents activate nerve cells in a cascade of neural processing that ultimately provides information about the immediate environment—information that is essential to navigation, finding food, finding mates. In short, information that is essential to survival.

The sensory receptors are an important key to our understanding how all senses operate—both the exotic and the ordinary. Different receptors are selectively sensitive to different physical stimuli. We say that they are tuned to different types of stimuli. Since these stimuli reflect important aspects of the world around us, we could just as easily say that they are tuned to features of the environment itself. There are different levels of receptor selectivity. A coarse level of tuning confers modality specificity upon a receptor. Auditory hair cells do not respond to light, and photoreceptors do not respond to sounds. Clearly, the quality of environmental energy must be appropriate for each type of receptor. At a finer level, however, we discovered that different receptors are tuned to *quantitative* differences within a single modality. In the auditory systems of bats and dolphins, different frequencies stimulate different hair cells, and we said the hair cells are tuned to different auditory frequencies. We further saw how the demands of biosonar lead to particularly precise turning for exactly the same frequencies that are present in the sonar signal. The capabilities of the sonar receiver are well matched to the frequency characteristics of the sonar emitter. In electroreceptive fish, variations in the electric field produce different patterns of activity in electroreceptors that are strategically distributed along the fish's body. Here, too, the electroreceptors were tuned to frequencies that were present in the electric discharge organ of that species. In this way, interference between species is minimized. Many different species can occupy the same stream without jamming each other's signals because each uses a different frequency band. To take a final example, the orientation of photopigments within different photoreceptors in a specialized part of the compound eye of honeybees permits them to detect the pattern of polarization in skylight, which in turn permits them to determine the position of the sun, even when the sun is not directly visible.

Receptor tuning is essential because the discriminative capacities of any sensory system rely on the degree to which qualitative and quantitative differences in stimuli produce different patterns of responses in arrays of receptor cells. If two different stimuli produce the exact same pattern of activity in a population of sensory receptors, no amount of neural processing of those receptor responses will permit discrimination of the signals. Another receptor system might be evolved that would readily appreciate the difference. We've certainly encountered vivid examples of this fundamental fact.

Usually, the stimulus selectivity of sensory receptors depends on the presence of some sort of anatomical specialization. For instance, the arrangement of cells within electroreceptors acts like a frequency filter: the cells that line the cavity, their adhesions to one other, and the cellular matrix within the pore all contribute to the receptor tuning. These accessory structures help ensure that the receptors remain selectively sensitive to those frequencies that are most relevant to each individual fish's survival—and during development, they can even achieve tuning properties that are uniquely matched to each individual's electric discharge.

The photoreceptors of honeybees provided perhaps the clearest illustration of the relationship between the anatomy of a receptor and its functional tuning. Most of the ommatidia are twisted along their length. This twisting results in a loss of polarization sensitivity, which for many visual functions is just fine. Human photoreceptors are insensitive to the plane of polarized light, a fact that does not seem to present us with any particular problem. But the honeybees found a use for polarized light. They evolved a specialized part of the eye that contains receptors that are not twisted, and it is this part, and this part alone, that is responsible for their E-vector compass.

In the domain of hearing, we found that the overall structure of the cochlea is very similar in all mammals. In each case, the variations in air pressure that we call sounds induced movement in a small membrane called the basilar membrane. Movement of the basilar membrane produces shearing forces on auditory hair cell receptors, which causes ionic currents to flow within the hair cells and begins the process of hearing. Yet different mammals are sensitive to different frequencies of acoustic energy. Variations in size and elasticity of the middle ear bones, and especially in the mechanical properties of the basilar membrane within the cochlea, is the reason why some species can hear sound frequencies that are completely inaudible to others.

In some systems, we have not yet identified the nature, or even the location, of the relevant receptor cells. This of course is a major obstacle in our attempts to further understand the sensory modality of magneto-reception.

While an understanding of the receptors is an essential element to understanding the mechanisms of sensations of any sort, this is just the beginning of the process. Naturally, if the *pattern* of responses within an array of sensory receptors is going to encode different aspects of our environment, some mechanisms must be able to preserve that coded information—and to use it to control other processes, especially behavior. This of course is the job of the central nervous system. We've seen how senses cannot possibly be understood without also understanding how the nervous system processes the information conveyed by receptor systems.

Sensory and perceptual systems are thus an elegant combination of specialized receptor cells with a complex network of nerve cells. Before we could make any progress in understanding how these systems work, we first had to develop a pretty clear and specific view of their purpose. The human visual system is a remarkably acute, general-purpose system. It provides us with a "high-fidelity" representation of shape, texture, color, and relative location of all objects that reflect light in the range from 400 to 700 nanometers. Such a general-purpose system is particularly complex and, as a result, particularly hard to understand. Many of the systems we have considered are easier to appreciate, in part because they may not be as general-purpose as the human visual system. They perform important tasks, to be sure. Many of these exotic systems are used to avoid obstacles, aid navigation, and find food. Pretty essential jobs. But jobs that don't necessarily provide the rich representation of the world that human vision does. Surely it is hard to see how any creature that can't avoid obstacles, move from place to place, and find food can possibly survive in this world. But if the task of a sensory system is restricted to those few essential functions, it might provide a good model for developing a deeper level of understanding applicable to many senses. In many instances, these exotic systems serve as models of biological sensory systems more generally.

This is why so many talented people study these systems. The hope is that they are simpler than our own sensory systems, because they dedicated to performing more specific tasks. Once we have an appreciation of the forms of stimulus energy a sensory system depends upon, and we have developed an idea of the ways in which that system is employed by its possessor, we can begin to formulate ideas about how that system might actually perform its task. Here is where we encounter computational issues, and as David Marr (an influential computer scientist who studied visual perception) so insightfully pointed out, we cannot hope to understand how complicated nervous systems perform their tasks without a clear idea of what those tasks actually are.

Early in our explorations, we touched on some of the major computational issues that arise in the use of a biosonar system. The goal of the system is to determine the distance, approach velocity, location, and perhaps the identity of objects by their echoes. The delay between the call and the echo determines target range, Doppler shifts of the echo specify approach velocity, and comparisons between the echoes arriving at each ear vary with target eccentricity. All the beautiful work on the biosonar system of the bat would not have been possible without an appreciation of the computational issues inherent in the implementation of biosonar. It is indeed gratifying that the different areas within the neocortex of the bat appear to be specifically related to processing approach velocity, target range, and the like. We take this as tentative confirmation of the validity of the computational approach, since the brain seems to have mechanisms devoted to the computations we identified.

Computational problems are invariably informed by knowledge of the constraints provided by physics, and can often be effectively and concisely described by using techniques of mathematics and the engineering sciences. Yet we have also seen that biological systems frequently employ anatomical solutions to computational problems. Thus, instead of having a neural network perform the equivalent of geometric calculations to determine the position of the sun by using E-vectors in skylight, honeybees solve the problem anatomically; by having the polarization preferences of receptors match the celestial pattern of polarization. This is a very elegant example of what we termed "computational anatomy." Other computational solutions might well be anatomically mediated. The bat's sonar system might provide another example, as might the jamming avoidance response of weakly electric fish. The myriad interconnected neurons in the vertebrate brain render these latter issues a very tough nut to crack, but the research continues, and so will the insights.

When considered from the behavioral, computational, and physiological perspectives, we see that these exotic senses are fundamentally similar to our own senses of vision, hearing, touch, taste, and smell. There is a unity in different sensory systems that we could not have anticipated nor understood without that combination of human intellect, ingenuity, and perseverance that we call science. The treatment of all senses as mechanisms provides us with that unity of understanding. They are wonderful mechanisms to be sure, but they are mechanisms. This does not diminish them, or us. For me, there is a beauty in all of this that far surpasses any explanation that might be born from speculation of even the most fertile imagination.

We started this journey by considering appropriate criteria for explanations of natural phenomena. It is appropriate, then, that we remind ourselves now that the explanations we have explored remain incomplete. There is still much to be done, much to be discovered.

Consider, for instance, the star-nosed mole, a creature that has captured the attention of Jon Kaas and Kenneth Catania of Vanderbilt University. A picture will reveal why this little creature is called the star-nosed mole. (See figure 20.1.)

Have you ever seen a nose like that? Here we have a proboscis only an astronomer could appreciate. And on the remote chance you're not too impressed with the appearance of that nose, consider this—those fingerlike appendages have little or nothing to do with the sense of smell! In this case, that elaborate snout is part of an exquisitely fine sense of touch. Sir Hiram would be pleased. At last we find a creature whose face is, in fact, covered with "sensitive spots."

These little creatures live in meandering tunnels underground. They hardly ever come to the surface, which probably explains why so few people know about them. Kenneth Catania has photographed these

Figure 20.1
A star-nosed mole making a rare aboveground appearance.

strange beasts as they search for food. They like to eat earthworms (figure 20.2). As described by Catania and Kaas, those little "nose fingers" are in constant motion as the mole searches for food. They are swept back when the animal raises its nose, and are thrust forward as the nose makes contact with the ground. These exploratory nose movements are made 10 or more times every second. As soon as one of the "fingers" comes in contact with prey, the mole quickly grabs it with its mouth and eats it. Interestingly, it has been suggested that the "nose fingers" of the star-nosed mole might be electroreceptive, but thus far Catania and Kaas have seen little to support that speculation.

The appearance of the claws doesn't suggest a great deal of dexterity or tactile sensitivity. Those claws appear to be well-suited for digging, but little else. But the nose fingers are something else altogether. They are very mobile, and they have a remarkably dense innervation. A close-up of these amazing little appendages appears in figure 20.3. Even though the nose is only about 1 cm across (about the size of one of your smallest fingernails), it is innervated by five times as many sensory fibers as your entire hand. One might suspect that such a magnitude of neural investment might produce a very sensitive organ indeed. Based on anatomical considerations, Catania and Kaas estimated that the nose fingers of the star-nosed mole might be able to detect variations in surfaces that are less than 1 millionth of a meter in magnitude. This is better than you can do with your fingertips, much less the end of your nose.

Figure 20.2
A star-nosed mole enjoying a light snack.

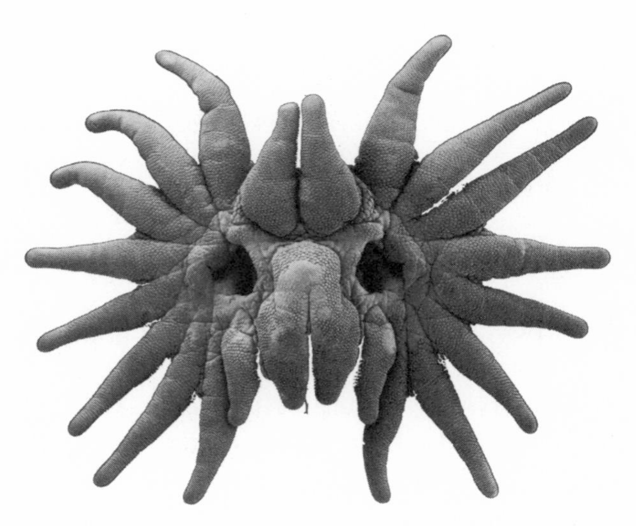

Figure 20.3
Scanning electron micrograph of the nasal appendages of the star-nosed mole.

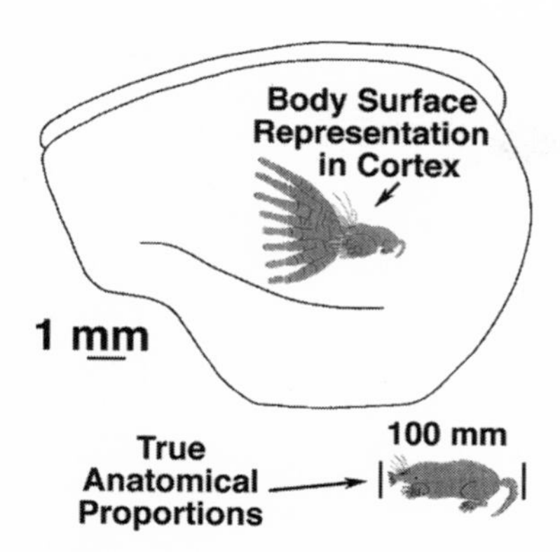

Figure 20.4
Magnification of the cortical representation of the mole's nose in the somatosensory cortex.

Another general characteristic of sensory systems is that the neural centers responsible for each sensory modality contain a map of the receptor surface. Spatial differences are often represented spatially within the brain. In vision, we have a map of the retina; in hearing, a map of the basilar membrane; in touch, a map of the skin surface. Stimulation of different receptors stimulates different neurons in the brain. In part, at least, that is why you can distinguish between being touched on your index finger from being touched on your thumb, even when your eyes are closed. As a rule, these neural maps of our skin are distorted—more brain tissue is devoted to the skin of the hand than to the skin of your back, even though your back is much larger than your hand. These distortions have a great deal to do with the relative sensitivity of different regions of the skin surface. For example, the acuity of your fingertips is much greater than the acuity of the skin on your back. Nowhere is this relationship between the acuity of different areas of skin and the size of the brain's map of that corresponding region more evident than in the brain of the star-nosed mole (figure 20.4). This distortion of the brain's representation of a receptor surface may sound familiar, because we encountered the exact same principle in the auditory system of the mustache bat. In that case, it was the representation of 60kHz—the second harmonic of the bat's sonar call—that received a disproportionately large representation within the brain.

So again we see that the exotic, highly specialized sensory/perceptual systems evolved by some animals represent variations of the same theme—they do not represent radically different themes.

One remaining issue concerns the unification of different sensory modalities. Certainly our subjective impression is of a unified perceptual event. The sound of friendly laughter is accompanied by a smiling face. Spoken words are closely associated with distinctive movements of the mouth, making lipreading possible. The events we perceive often produce energy in multiple forms, and while those forms may initially be encoded by modality-specific sensory systems, they appear ultimately to gain access to common neural circuits. It is this convergence of the senses onto common neural circuits that is thought to underlie the phenomenological unity of the perceptual world.

The same principles of polymodal integration are apparent in more exotic systems. For example, certain species of snakes are endowed with small pits distributed along the snout. These pits contain sensory receptors that respond to infrared radiation—that is, heat. Information from these thermal receptors is conveyed to the central nervous system of the snake, and in at least one neural structure, called the optic tectum, it is combined with visual information. Within the optic tectum there are neurons that respond to visual inputs, infrared inputs, or both visual and infrared signals. Any given cell within the optic tectum responds only to stimuli located within a small region of space. Neurons that receive both infrared and visual inputs are called bimodal neurons, and the bimodal neurons in the optic tectum of snakes require that the infrared and visual signals originate from the same spatial location. As a result, these cells are most vigorously stimulated by small, warm objects that are visible. Since different neurons sample different spatial locations, a small, warm, visible object in front of the snake will always stimulate some cells within the optic tectum—which neurons are active depends on the exact location of the object. The optic tectum in turn controls neurons that control orienting of the head. The result? A rapid strike, and an unsuspecting mouse becomes a meal.

It turns out that every vertebrate—including humans—has a structure that appears analogous to the optic tectum. And in every case, this structure receives inputs from multiple sensory modalities, and can issue commands that produce orienting responses. In electric fish, the optic tectum is a site of convergence between the visual system and the electroreceptive system. In primates, an analogous structure controls eye movements, and receives converging inputs from the visual and auditory systems. In this way, we can orient our gaze toward visual events, auditory events, and—especially—visual/auditory events. Again we have a vivid example of the fact that all sensory modalities share similar design features: what we learn about one often applies to others.

We often find multiple, mutually reinforcing systems operating in the service of essential tasks. This was certainly the case in birds, where the sun, stars, and geomagnetic fields all provide important information for navigating during long migrations. This represents a form of multimodal integration that has proved to be very complicated and difficult to study Untangling the complicated interactions between sensory modalities is surely one of the difficult problems facing the next generation of sensory psychologists, zoologists, and neurobiologists. Consider, for instance, the remarkable ability of spawning salmon to return to the exact stream where they hatched. We know from tagging studies that salmon find their way back to the right stream, of the right tributary, of the right river, after being at sea for four years or more. How do they do it? It appears that salmon have a magnetoreceptive system, but it seems incomprehensible that magnetoreceptive cues alone could provide them with the information needed to return to their natal stream after such a long absence. One possibility is that they can distinguish the water of their natal stream by using their sense of smell. But they would have to remember that smell for a long time. There is speculation that early exposure to the odor of the stream could influence the development of the fish's nervous system—perhaps using a process analogous to imprinting in birds.

We are obviously a long way from understanding these and other mysteries of nature. We don't yet have all the answers, but at least we are learning how to ask the right questions. The noted British naturalist Thomas Huxley observed that taking a stroll on the beach, or a hike through the woods, without the benefit of a knowledge of natural science was like walking through a magnificent art museum in which most of the paintings face the wall. Perhaps we have done nothing more than gain a brief glimpse of some of nature's more surrealistic masterpieces. Still, it is better to have enjoyed a glimpse then to not have known them at all.

Source Notes

In this section, I list the original sources of the illustrations. In many cases, the illustrations were drawn by me and are therefore not listed below.

Figure 1.1 From http://web.psych.ualberta.ca/~medler/courses/psyco281/lecture01/slide4.html; downloaded on April 9, 1998.

Figure 2.1 From Sir Hiram Maxim, "The Sixth Sense of the Bat," *Scientific American,* supp. (Sept. 7, 1912): 148.

Figure 2.2 Maxim, "The Sixth Sense of the Bat," 149.

Figure 2.3 Maxim, "The Sixth Sense of the Bat," 149.

Figure 2.4 Maxim, "The Sixth Sense of the Bat," 148.

Figure 2.5 Photograph from Merlin D. Tuttle, *America's Neighborhood Bats* (Austin: University of Texas Press, 1988), p. 7.

Figure 2.6 Photographs by Merlin D. Tuttle, "Saving North America's Beleaguered Bats," *National Geographic* (August 1995): 42.

Figure 2.7 Photographs from Tuttle, *America's Neighborhood Bats,* pp. 6, 7.

Figure 3.6 G. D. Pollack and J. H. Casseday, "The Neural Basis of Echolocation in Bats," *Zoophysiology* 25 (1989): 10, fig.1.2.

Figure 3.7 Photo obtained from www.warplane.org., March 1, 1998.

Figure 4.1 Nubuo Suga, "Biosonar and Neural Computation in Bats," *Scientific American* 262 (6) (June 1990): 62.

Figure 4.3 Drawing from Moïse H. Goldstein, "The Auditory Periphery," in Vernon B. Mountcastle (ed.), *Mountcastle's Medical Physiology,* (St. Louis: C. V. Mosby Co., 1974), 383, fig. 12-2a. Artwork by Carly Hughes.

Figure 4.4 N. R. Carlson, *The Physiology of Behavior* (Boston: Allyn & Bacon, 1980), p. 188, fig. 8.12. Artwork by Carly Hughes.

Figure 4.5 Goldstein, "The Auditory Periphery," p. 383, fig. 12–2b (A); p. 391, fig. 12-9 (B); p. 405, fig. 12-27 (C). Artwork by Carly Hughes.

Figure 4.8 Pollack and Casseday, "The Neural Basis of Echolocation in Bats," *Zoophysiology* 25 (1989): 38, fig. 2.14.

Figure 4.9 Suga, "Biosonar and Neural Computation in Bats," 60.

Figure 4.12 Adapted from Suga, "Biosonar and Neural Computation in Bats," 64.

Figure 4.14 Adapted from Suga, "Biosonar and Neural Computation in Bats," 63.

Figure 4.15 Adapted from Suga, "Biosonar and Neural Computation in Bats," 64.

Figure 4.16 H.-U. Schnitzler et al., "The Acoustic Image of Fluttering Insects in Echolocating Bats," in F. Huber and H. Markl (eds.), *Neuroethology and Behavioral Physiology: Roots and Growing Pains* (Berlin: Springer-Verlag, 1983), p. 235.

Figure 5.1 Photograph from Whitlow W. Au, *The Sonar of Dolphins* (New York: SpringerVerlag, 1993), fig. 1.1.

Figure 5.2 Ashley Montagu and John C. Lilly reproduced G. Elliot Smith's 1902 drawing in *The Dolphin in History* (Los Angeles: William Andrews Clark Memorial Library, UCLA, 1963).

Figure 5.3 Photo from John E. Heyning, *Masters of the Ocean Realm: Whales, Dolphins & Porpoises* (Seattle: University of Washington Press, 1995), p. 24.

Figure 6.1 Adapted from P. E. Purves and G. E. Pilleri, *Echolocation in Whales and Dolphins* (New York: Academic Press, 1983), fig. 13.

Figure 6.2 Au, *The Sonar of Dolphins,* fig. 5.14.

Figure 6.3 Au, *The Sonar of Dolphins,* fig. 5.16.

Figure 6.4 Au, *The Sonar of Dolphins,* fig. 5.3.

Figure 6.5 Au, *The Sonar of Dolphins,* figs. 5.3 and 5.4.

Figure 7.1 Au, *The Sonar of Dolphins,* fig. 2.4. Redrawn by Kelly Hughes.

Figure 8.1 the program of the 36th Annual meeting of the Psychonomic Society, 1995.

Figure 8.2 Map by Matthew Paris, original in the British Museum. Taken from Leo Bagrow, *History of Cartography,* rev. and enl. by R. A. Skelton, 2nd ed. (Chicago: Precendent Publishing, Inc., 1985), color plate B.

Figure 8.3 L. Humphreys, *Antique Maps and Charts* (New York: Dorset Press, 1989), p. 147.

Figure 9.1 R. Robin Baker, *Bird Navigation: The Solution to a Mystery?* (New York: Holmes and Meier, 1984), p. 30, fig. 3.7.

Figure 9.2 K. Schmidt-Koenig, J. U. Ganzhorn, and R. Ranvaud, "The Sun Compass," in P. Berthold (ed.), *Orientation in Birds* (Birkhauser Verlag, 1991). p. 2, fig. 1.

Figure 9.3 W. Wiltschko and R. Wiltschko, "Magnetic Orientation and Celestial Cues in Migratory Orientation," in Berthold (ed.), *Orientation in Birds,* p. 21, fig. 4.

Figure 9.7 Adapted from Baker, *Bird Navigation,* figs. 8.1 and 8.2.

Figure 9.9 Baker, *Bird Navigation,* fig. 8.5.

Figure 10.1 Talbot H. Waterman, *Animal Navigation* (New York: Scientific American Books, 1989), p. 165.

Figure 10.2 M. M. Walker et al., "Structure and Function of the Vertebrate Magnetic Sense," Nature 390: 375, fig. 6.

Figure 10.3 Part A is from *Les Oeuvres de Descartes,* vol. 11, *Le Monde, Description du Corps Humain, Passions de l'Ame, Anatomica Varia* (Paris: Librairie Philosophique J. Vrin, 1986). Part B is from S. J. de Armond, M. M. Fucso, and M. M. Dewey, *A Photographic Atlas: Structure of the Human Brain,* 2nd ed. (New York: Oxford University Press, 1976), p. 9.

Figure 11.6 Photographs from Karl von Frisch, *The Dance Language and Orientation of Bees,* (Cambridge, Mass.: The Belknap Press of Harvard University Press, 1967), p. 485, fig. 421.

Figure 11.7 Photographs from Frisch, *The Dance Language and Orientation of Bees,* p. 486, fig. 423.

Figure 11.9 Adapted from Eugene Hecht, *Optics* (Reading, Mass.: Addison-Wesley, 1987), fig. 8.10.

Figure 11.10 Adapted from Eugene Hecht, *Optics* (Reading, Mass.: Addison-Wesley, 1987), fig. 8.37.

Figure 11.11 Waterman, *Animal Navigation,* p. 113.

Figure 11.12 Part A is from H. Davson (ed.), *The Eye,* vol. 1, *Vegetative Physiology and Biochemistry* (New York: Academic Press, 1962), p. 2, fig. 1. Part B is from Davson (ed.), *The Eye,* vol. 1, p. 5, fig. 3. C is from H. Davson (ed.), *The Eye,* vol. 2, *The Visual Process* (New York: Academic Press, 1962), p. 14, figs. 1 and 2. D is from R. L. and K. K. de Valois, *Spatial Vision* (Oxford: Oxford University Press, 1990), p. 61, fig. 3.5. E is from J. L and C. G Gould, *The Honey Bee* (New York: Scientific American Library, 1988), p. 80. Parts F and G are from M. L. Winston,*The Biology of the Honey Bee* (Cambridge, Mass.: Harvard University Press, 1987), p. 16, fig. 3.3. H is from M. Rothschild, Y. Schlein, and S. Ito, *A Colour Atlas of Insect Tissues via the Flea* (Weert, The Netherlands: Wolfe Publishing, 1986), p. 168, fig. 2.

Figure 11.13 Parts A and B are from J. L. Gould and C. G. Gould, *The Honey Bee* (New York: Scientific American Library, 1988), p. 43. C is from K. Kirschfeld, "The Resolution of Lens and Compound Eyes," in F. Zettler and R. Weiler (eds.), *Neural Principles in Vision* (Berlin: Springer-Verlag, 1976), p. 365, fig. 7C.

Figure 11.14 R. Wehner, "Polarized-Light Navigation by Insects," *Scientific American* 235 (1976): 106–115.

Figure 11.15 R. Wehner, "Polarized-Light Navigation by Insects," 106–115.

Figure 11.16 R. Wehner and S. Strasser, "The POL Area of the Honey Bee's Eye: Behavioral Evidence," *Physiological Entomology* 10 (1985): 338, fig. 1.

Figure 11.17 R. Wehner and Strasser, "The POL Area of the Honey Bee's Eye: Behavioral Evidence," 340, fig. 2C.

Figure 11.18 S. Rossel and R. Wehner, "The Bee's E-Vector Compass," in R. Menzel and A. Mercer (eds.), *Neurobiology and Behavior of Honeybees,* (Berlin: Springer-Verlag, 1987), p. 83, fig. 6.

Figure 12.1 http://www.snap-shot.com/photos/fish/sharks/tiger/tiger24. htm (June 22, 1998). Photo by Donovan Gutierrez.

Figure 12.2 Wu, "Electric Fish and the Discovery of Animal Electricity," *American Scientist* 72 (1984): 600, fig. 2.

Figure 12.3 Wu, "Electric Fish and the Discovery of Animal Electricity," 598–607. Wu got it from T. H. Feder, *Great Treasures of Pompeii and Herculaneum* (New York: Abbeville, 1978).

Figure 12.4 http://phylogeny.arizona.edu/tree/eukaryotes/animals/chordata/actinopterygii/gymnotiformes/gymnotiformes.html.

Figure 12.5 From ibid.

Figure 12.7 H. W. Lissman and K. E. Machin, "The Mechanism of Object Location in *Gymnarchus Niloticus* and Similar Fish," *The Journal of Experimental Biology* 35 (1958): 451–486.

Figure 13.1 Part A is from http://weber.u.washington.edu/~islander/fish.html. Part B is from http://www.dinofish.com/image13.htm.

Figure 13.2 S. Dijkgraaf, "The Functioning and Significance of the Lateral-Line Organs," *Biological Reviews* 38 (1962): 54, fig. 1.

Figure 13.3 Dijkgraaf, "The Functioning and Significance of the Lateral-Line Organs," p. 76, fig. 13.

Figure 13.4 Part A is from D. Bodznick and R. L. Boord, "Electroreception in Chondrichthyes: Central Anatomy and Physiology," in T. H. Bullock and W. Heiligenberg (eds.), *Electroreception* (New York: John Wiley & Sons, 1986), p. 231, fig. 1. Part B is from A. J. Kalmijn, "Electric and Magnetic Sensory World of Sharks, Skates, and Rays," in E. S. Hodgson and R. F. Mathewson (eds.), *Sensory Biology of Sharks, Skates, and Rays* (Arlington, Va.: Office of Naval Research, 1978), p. 509, fig. 1.

Figure 13.5 M. V. L. Bennett and S. Obara, "Ionic Mechanisms and Pharmacology of Electroreceptors," in Bullock and Heiligenberg (eds.), *Electroreception,* p. 166, fig. 6.

Figure 14.1 Both parts A and B are from H. H. Zakon, "The Electroreceptive Periphery," in Bullock and Heiligenberg (eds.), *Electroreception,* p. 107, fig. 1 (A), and p. 122, fig. 4 (B).

Figure 14.2 Lissman and Machin, "The Mechanism of Object Location in *Gymnarchus Niloticus* and Similar Fish," p. 456, fig. 2.

Figure 15.1 Adapted from J. C. Dye and J. H. Meyer, "Central Control of the Electric Organ Discharge in Weakly Electric Fish," in Bullock and Heiligenberg (eds.), *Electroreception,* pp. 74–75, fig. 1.

Figure 15.3 Adapted from Dye and Meyer, "Central Control of the Electric Organ Discharge in Weakly Electric Fish," in Bullock and Heiligenberg (eds.), *Electroreception,* p. 79, fig. 3.

Figure 15.4 Adapted from Dye and Meyer, "Central Control of the Electric Organ Discharge in Weakly Electric Fish," in Bullock and Heiligenberg (eds.), *Electroreception,* p. 81, fig. 6.

Figure 16.1 Adapted from A. H. Bass, "Electric Organs Revisited," in Bullock and Heiligenberg (eds.), *Electroreception,* p. 16, fig. 1.

Figure 16.2 Adapted from Dye and Meyer, "Central Control of the Electric Organ Discharge in Weakly Electric Fish," in Bullock and Heiligenberg (eds.), *Electroreception,* pp. 74–75, fig. 1.

Figure 16.3 Adapted from Dye and Meyer, "Central Control of the Electric Organ Discharge in Weakly Electric Fish," in Bullock and Heiligenberg (eds.), *Electroreception,* pp. 74–75, fig. 1.

Figure 16.4 Adapted from Dye and Meyer, "Central Control of the Electric Organ Discharge in Weakly Electric Fish," in Bullock and Heiligenberg (eds.), *Electroreception,* pp. 74–75, fig. 1.

Figure 16.5 Zakon, "The Electroreceptive Periphery," in Bullock and Heiligenberg (eds.), *Electroreception,* p. 134, fig. 7B.

Figure 16.6 C. D. Hopkins, "Electric Communication in Fish," *American Scientist* 62 (1974): 431, fig. 6.

Figure 17.1 Photo from Theodore Rowland-Entwistle, "The World You Never See. Insect Life," in conjunction with Oxford Scientific Films Ltd. (Chicago: Rand McNally & Company, 1976), p. 33, fig. 64.

Figure 17.2 Photo from Maurice and Robert Burton, *Encyclopedia of Insects and Arachnids* (New York: Crown Publishers, 1975), p. 149.

Figure 18.1 Carlson, *The Physiology of Behavior,* p. 292, fig. 9.13.

Figure 18.3 de Armond, Fucso, and Dewey, *A Photographic Atlas: Structure of the Human Brain,* p. 7

Figure 18.4 Adapted from Carlson, *The Physiology of Behavior,* p. 199, fig. 6.30.

Figure 19.1 Photo from D. Michael Stoddart, *The Scented Ape* (Cambridge, UK: Cambridge University Press, 1990), p. 59, fig. 3.4.

Figure 20.1 Photograph courtesy of Dr. Kenneth Catania, Vanderbilt University.

Figure 20.2 Photograph courtesy of Dr. Kenneth Catania, Vanderbilt University.

Figure 20.3 Modified from K. C. Catania et al., "Nose Stars and Stripes," *Nature* 364 (1993): 493. Micrograph courtesy of Dr. Kenneth Catania, Vanderbilt University.

Figure 20.4 Modified from K. C. Catania and J. H. Kaas, "The Unusual Nose and Brain of the Star-Nosed Mole," *BioScience* 46: 581, 584, figs. 5, 9.

The following list itemizes the sources for the illustrations used at the beginning of each chapter.

Chapter 1 Glover Morril Allen, *Bats* (New York: Dover Allen, 1967), p. 23, fig. 5.

Chapter 2 Ronald M. Nowack, *Walker's Bats of the World* (Baltimore: Johns Hopkins, 1994), p. ii.

Chapter 3 Tuttle, "Saving North America's Beleaguered Bats," 43.

Chapter 4 Suga, "Biosonar and Neural Computation in Bats," 64.

Chapter 5 Jacques-Yves Cousteau and Philippe Diole, *Dolphins* (Garden City, NY: Doubleday, 1975), p. 240.

Chapter 6 Adapted from Waterman, *Animal Navigation,* p. 132.

Chapter 7 Au, *The Sonar of Dolphins,* fig. 2.4.

Chapter 8 Robert Burton, *Bird Migration* (London: Eddison Sadd Editions, 1992), p. 18.

Chapter 9 Burton, *Bird Migration*, p. 135.

Chapter 10 Adapted from Waterman, *Animal Navigation,* p. 165.

Chapter 11 The first drawing of the compound eye of a honey bee, which was drawn by Jan Swammerdam in 1737. Reprinted in Rüdiger Wehner, *The Handbook of Sensory Physiology,* vol. VII/6C, *Vision in Invertebrates* (Berlin: Springer-Verlag, 1981), fig. 1.

Chapter 12 Adapted from Waterman, *Animal Navigation,* p. 153.

Chapter 13 Theodore H. Bullock and Walter Heiligenberg, eds., *Electroreception* (New York: Wiley, 1986). Fish is from Mark Ronan, "Electroreception in Cyclostomes," p. 215, fig. 3, and electroreceptors in the background are from Zakon, "The Electroreceptive Periphery," p. 143, fig. 10.

Chapter 14 Adapted from Bodznick and Boord, "Electroreception in Chondrichthyes: Central Anatomy and Physiology," p. 231, fig. 1.

Chapter 15 Adapted from Bass, "Electric Organs Revisited," p. 16, fig. 1.

Chapter 16 Adapted from Mary Hagedorn, "The Ecology, Courtship, and Mating of Gymnotiform Electric Fish," in Bullock and Heiligenberg, eds., *Electroreception,* p. 520, fig. 9.

Chapter 17 Mark Young, *The Natural History of Moths* (London: T. & A. D. Poyser, 1997), p. 167.

Chapter 18 The rat is from S. A. Barnett, *The Rat: Study in Behavior* (Chicago: University of Chicago Press, 1975); molecule from S. J. Schindler's cover illustration, *Angewandte Chemie* 33, no. 19 (1994).

Chapter 19 Anthony A. Pearson, "The Development of the *Nervus Terminalis* in Man," *The Journal of Comparative Neurology* 75 (1941): 39–66.

Chapter 20 Courtesy of Dr. Kenneth Catania; modified from Catania et al., "Nose Stars and Stripes," 493.

References

Able, K. P. (1994). Magnetic orientation and magnetoreception in birds. *Progress in Neurobiology* 42: 449–473.

Altes, R. A., and Anderson, G. M. (1980). "Binaural estimation of cross-range velocity and optimum escape maneuvers by moths." In R.-G. Busnel and J. F. Fish, eds., *Animal Sonar Systems* pp. 851–852. New York: Plenum.

Au, Whitlow W. L. (1993). *The Sonar of Dolphins*. New York: Springer-Verlag.

Baker, R. (1984). *Bird Navigation: The Solution to a Mystery?* New York: Holmes and Meier Publications.

Baker, R. (1985). "Magnetoreception by man and other primates." In J. L. Kirschvink, D. S. Jones, and B. J. MacFadden, eds. *Magnetic Biomineralization and Magnetoreception in Organisms*, pp. 537–561. New York: Plenum Press.

Bass, A. H. (1986). "Electric organs revisited." In T. H. Bullock and W. Heiligenberg (eds.), *Electroreception*, pp. 13–70. New York: John Wiley and Sons.

Bastian, J. (1986). "Electrolocation: behavior, anatomy and physiology." In T. H. Bullock and W. Heiligenberg, eds., *Electroreception*, pp. 577–612. New York: John Wiley and Sons.

Beauchamp, G. K., Yamazaki, K., and Boyse, E. A. (1985). The chemosensory recognition of genetic individuality. *Scientific American* 253(1): 86–92.

Bennett, M. V. L. (1971). "Electric organs." In W. S. Hoar and D. J. Randall, eds., *Fish Physiology* vol. 5, pp. 493–574. New York: Academic Press.

Bennett, M. V. L., and Obara, S. (1986). "Ionic mechanisms and pharmacology of electroreceptors." In T. H. Bullock and W. Heiligenberg, eds., *Electroreception*, pp. 157–181. New York: John Wiley and Sons.

Bennett, M. V. L., and Steinbach, A. B. (1969). "Influence of electric organ control system on electrosensory afferent pathways in mormyrids." In R. Llinas, ed., *Neurobiology of Cerebellar Evolution and Development*. Chicago: American Medical Association.

Benton, D. and Wastell, V. (1986). Effects of androstenol on human sexual arousal. *Biological Psychology* 22: 141–147.

Berliner, D. L., Monti-Bloch, L., Jennings-White, C., and Diaz-Sanchez, V. (1996). The functionality of the human vomeronasal organ (VNO): Evidence for steroid receptors. *Journal of Steroid Biochemistry and Molecular Biology,* 58: 259–265.

Blakemore, R. P. (1975). Magnetotaxic bacteria. *Science* 190: 377–379.

Brill, R. L., Sevenich, M. L., Sullivan, . J., Justman, J. D., and Witt, R. E. (1988). Behavioral evidence for hearing through the lower jaw by an echolocating dolphin (Tursiops truncatus). *Marine Mammal Science* 4: 223–230.

Bullock, T. H. (1982). Electroreception. *Annual Review of Neuroscience* 5: 121–170.

Bullock, T. H., Grinnell, A. D., Ikezono, E., Kameda, K., Katsuki, Y., Nomoto, M., Sato, O., Suga, N., and Yanagisawa, K. (1968). Electrophysiological studies of the central auditory mechanisms in cetaceans. *Zeitschrift für Vergleichende Phys.* 59: 117–316.

Catania, Kenneth C., and Jon H. Kaas. (1996). The unusual nose and brain of the star-nosed mole. *BioScience*, 46, no. 8: 578–586.

Cowley, J. J., Johnson, A. L., and Brooksbank, B. W. L. (1977). The effect of two odorous compounds on performance in an assessment-of-people test. *Psychoneuroendocrinology* 2: 159–172.

Darwin, C. (1859). *The Origin of Species by Means of Natural Selection or, the Preservation of Favored Races in the Struggle for Life*, 6th ed., pp. 178–179. A. L. Burt.

Daves, G. D., Lee, T. D., Rasmussen, L. E. L., Roelofs, W. L., and Zhang, A. (1996). Insect pheromones in elephants. *Nature* 379: 684–684.

Dijkgraaf, S. (1960). Hearing in bony fishes. *Proceedings of the Royal Society of London* 152: 51–54.

Dijkgraaf, S. (1963). The functioning and significance of the lateral-line organs. *Biological Reviews* 38: 51–105.

Dye, J. C., and Meyer, J. H. (1986). "Central control of the electric organ discharge in weakly electric fish." In T. H. Bullock and W. Heiligenberg, eds., *Electroreception*, pp. 71–103. New York: John Wiley and Sons.

Faraday, M. (1832). Experimental researches in electricity. *Philosophical Transactions of the Royal Society of London* 122(1): 124–194.

Fillion, T. J., and Blass, E. M. (1986). Infantile experience with suckling odors determines adult sexual behavior in male rats. *Science* 231: 729–731.

Filsinger, E. E., Braun, J. J., Monte, W. C., and Linder, D. E. (1984). Human (*Homo sapiens*) responses to the pig (*Sus scrofa*) sex pheromone 5 Alpha-androst-16-en-3-one. *Journal of Comparative Psychology* 98: 219–222.

Frisch, Karl von. (1967). *The Dance Language and Orientation of Bees.* Cambridge, Mass.: Belknap Press of Harvard University Press.

Fromme, H. G. (1961). Untersuchungen uber das Orientierungsvermogen nachtlich ziehender Kleinvogel (*Erithacus rubecula, Sylvia communis*). *Zeits. Tierpsychol.* 18: 205–220.

Galambos, Robert. (1942). The avoidance of obstacles by flying bats: Spallanzani's ideas (1794) and later theories. *Isis* 34: 132–140.

Gould, J. L., and Gould, C. G. (1988). *The Honey Bee.* New York: Scientific American Library.

Grier, J. W., and Burk, T. (1992). *Biology of Animal Behavior.* St. Louis: Mosby Year Book.

Griffin, Donald R. (1986). *Listening in the Dark. The Acoustic Orientation of Bats and Men.* Ithaca, N.Y.: Cornell University Press.

Hagedorn, M., and Heiligenberg, W. (1985). Court and spark: Electric signals in the courtship and mating of gymnotoid electric fish. *Animal Behavior* 33: 254–265.

Halpern, M. (1987). The organization and function of the vomeronasal system. *Annual Review of Neuroscience* 10: 325–362.

Hecht, E. (1987). *Optics.* New York: Addison-Wesley.

Heiligenberg, W. (1986). "Jamming avoidance responses. In T. H. Bullock and W. Heiligenberg, eds., *Electroreception,* pp. 613–649.

Hopkins, C. D. (1988). Neuroethology of electric communication. *Annual Review of Neuroscience* 11: 497–535.

Kalmijn, A. J. (1966). Electro-perception in sharks and rays. *Nature* 212: 1232–1233.

Kalmijn, A. J. (1971). The electric sense of sharks and rays. *Journal of Experimental Biology* 55: 371–383.

Kalmijn, A. J. (1988). "Hydrodynamic and acoustic field detection." In J. Atema, R. R. Fay, A. N. Popper, and W. N. Tavolga, eds., *Sensory Biology of Aquatic Animals.* New York: Springer-Verlag.

Kellogg, W. N., and Kohler, R. (1952). Responses of the porpoise to ultrasonic frequencies. *Science* 116: 250–252.

Klingmuller, D., Dewes, W., Krahe, T., Brecht, G., and Schweibert, H. U. (1987). Magnetic resonance imaging of the brain in patients with inosmia and hypothalamic hypogonadism (Kallman's syndrome). *Journal of Clinical Endocrinology and Metabolism* 65: 581–584.

Labhart, T. (1980). Specialized photoreceptors at the dorsal rim area of the honeybee's compound eye: Polarizational and angular sensitivity. *Journ. of Comp. Physiol. A.* 141: 19–30.

Leask, M. J. M. (1977). A physiochemical mechanism for magnetic field detection by migrating birds and homing pigeons. *Nature* 267: 144–145.

Lissman, H. W. (1958). On the function and evolution of electric organs in fish. *Journal of Experimental Biology* 35(1): 156–192.

Lissman, H. W. (1963). Electric location by fishes. *Scientific American* 208(3): 50–59.

Lissman, H. W., and Machin, K. E. (1958). the mechanism of object location in Gymnarchus niloticus and similar fish. *Journal of Experimental Biology* 35(2): 451–486.

Lorenzini, S. (1678/1705). *Osservazioni intorno alle Torpedini* (The curious and accurate observations of Mr. Stephen Lorenzini of Florence), J. Davis (trans.), pp. 66–72. London: Jeffery Wale.

Malmström, V. H. (1976). Knowledge of magnetism in pre-Columbian mesoamerica. *Nature* 259: 390–391.

Marr, David. (1982). *Vision. A Computational Investigation into the Human Representation of Processing of Visual Information.* San Francisco: W. H. Freeman.

Maxim, Sir Hiram. (1912). The sixth sense of the bat. Sir Hiram Maxim's contention. The possible prevention of sea collisions. *Scientific American*, suppl. (Sept. 7, 1912): 148–151.

McBride, Arthur F. (1956). Evidence for echolocations by cetaceans. *Deep-Sea Research* 3: 153–154.

McClintock, M. K. (1971). Menstrual synchrony and suppression. *Nature* 229: 244–245.

Menzel, R., and A. Mercer, eds. (1987). *Neurobiology and Behavior of Honeybees.* Berlin: Springer-Verlag.

Merkel, F. W., and Fromme, H. G. (1958). Untersuchungen uber Orientierungsvermogen nachtlich Ziehender Rotkehlchen, *Erithacus rubecula. Naturwissen* 45: 499–500.

Middendorf, A. von. (1859). Die Isepipetsen Russlands: Grundlagen zur Erforschung der Zugzeiten und Zugrichtungen der Vogel Russlands. *Mem. Acad. Sci. St. Petersbourg* 8: 1–143.

Montgomery, J. (1984). Noise cancellation in the electrosensory system of the thornback ray: common mode rejection of input produced by the animal's own ventilatory movement. *Journal of Comparative Physiology* 155: 102–112.

Monti-Bloch, L., and Grosser, B. I. (1991). Effect of putative pheromones on the electrical activity of the human vomeronasal organ and olfactory epithelium. *Journal of Steroid Biochemistry and Molecular Biology* 39: 573–582.

Mulrine, Anna. (1997). Wish you had that nose? *U.S. News and World Report,* January 13, p. 55.

Newman, Eric A., and Hartline, Peter H. (1982). The infra-red "vision" of snakes. *Scientific American*, 246, no. 3: 117–127.

Norris, K. S. (1964). Some problems of echolocation in cetaceans. In W. N. Tavolga, ed., *Marine Bioacoustics*, pp. 316–336. New York: Pergamon Press.

Norris, K. S. and Harvey, G. W. (1974). Sound transmission in the porpoise head. *Journal of the Acoustical Society of America* 56: 659–664.

Parker, G. H. and van Heusen, A. P. (1917). The responses of the catfish, *Amiurus nebulosus,* to metallic and non-metallic rods. *American Journal of Physiology*, 44: 405–420.

Pierce, G. W. and Griffin, D. R. (1938). Experimental determination of supersonic notes emitted by bats. *Journal of Mammology* 19: 454–455.

Pollack, G. D., and J. H. Casseday. (1989). *The Neural Basis of Echolocation in Bats*. Berlin: Springer-Verlag.

Puttnam, Clare. (1993). Noisy moths leave a nasty taste in the mouth. *New Scientist*, no. 1876 (June 5): 16, 138.

Rasmussen, L. E. L., Lee, T. D., Roelofs, W. L., Zhang, A. L., and Daves, G. D. (1996). Insect pheromone in elephants. *Nature* 379: 684.

Roeder, K. D. (1967). *Nerve Cells and Insect Behavior.* Cambridge, Mass.: Harvard University Press.

Rossel, S. R., and Werner, R. (1987). The bee's E-vector compass, in R. Menzel, ed., *Honeybee Neurology,* pp. 76–93. New York: Springer-Verlag.

Rothschild, M., Y. Schlein, and S. Ito. (1986). *A Colour Atlas of Insect Tissues (Via the Flea).* Weert, Netherlands: Wolfe Publishing.

Russell, M. J., Switz, G. M., and Thompson, K. (1980). Olfactory influences on the human menstrual cycle. *Pharmacology, Biochemistry and Behavior* 13: 737–738.

Schnitzler, Hans-Ulrich, and Henson, O'Dell W. (1980). Performance of airborne animal sonar system: I. Microchirtoptera. In R.-G. Busnel and J. F. Fish, eds., *Animal Sonar Systems*, pp. 235–250. New York: Plenum Press.

Semm, P., and Demaine, C. (1986). Neurophysiological properties of magnetic cells in the visual system of the pigeon. *Journal of Comparative Physiological A.* 159: 619–625.

Semm, P., Nohr, D., Demaine, C. and Wiltschko, W. (1984). Neural basis of the magnetic compass: Interactions of visual, magnetic and vestibular inputs in the pigeon's brain. *Journal of Comparative Physiological A.* 155: 283–288.

Semm, P., Schneider, T., and Vollrath, L. (1980). Effects of an earth-stregth magnetic field on electrical activity in pineal cells. *Nature* 288: 607–608.

Stern, K. and McClintock, M. K. (1998). Regulation of ovulation by human pheromones. *Nature* 392: 177–179.

Stoddart, D. Michael (1990). *The Scented Ape.* Cambridge: Cambridge University Press.

Suga, Nubuo. (1990). Biosonar and neural computation in bats. *Scientific American* 262, no. 6 (June): 60–68.

Szabo, T. (1974). "Anatomy of the specialized lateral line organs of electroreception." In A. Fessard, ed., *The Handbook of Sensory Physiology*, vol. III/3, pp. 13–58. New York: Springer-Verlag.

Tipler, P. A. (1982). *Physics*. New York: Worth Publishers

Veith, J. L., Buck, M. Getzlaf, S., Vandalfsen, P. and Slade, S. (1983). Exposure to men influences the occurrence of ovulation in women. *Physiology and Behavior* 31: 313–315.

Walker, M., Diebel, C., Haugh, C., Pankhurst, P., Montgomery, J., and Green, C. (1997). Structure and function of the vertebrate magnetic sense. *Nature* 390: 371–376.

Watanabe, A., and Takeda, K. (1963). The change of discharge frequency by AC stimulation in a weak electric fish. *Journal of Experimental Biology* 40: 57–66.

Waterman, T. H. (1989). *Animal Navigation*. New York: Scientific American Library.

Wedekind, C., Seebeck, T., Bettens, F., and Paepcke, A. J. (1995). MHC-dependent mate preferences in humans. *Proceedings of the Royal Society of London*, B260: 245–249.

Wehner, R. (1976). Polarized-light navigation by insects. *Scientific American* 253: 106–115.

———. (1989). Neurobiology of polarization vision. *Trends in Neurosciences* 12: 353–359.

Wehner, R., and Strasser, S. (1985). The POL area of the honey bee's eye: Behavioural Evidence. *Physiological Entomology* 10: 337–349.

Whitman, C. O. (1899). Myths in animal psychology. *Monist* 9: 524–537.

Wiltschko, W., and Gwinner, E. (1974). Evidence for an innate magnetic compass in garden warblers. *Naturwissen* 61: 406.

Wiltschko, W., and Wiltschko, R. (1972). Magnetic compass of European robins. *Science* 176: 62–64.

Winston, M. L. (1987). *The Biology of the Honey Bee*. Cambridge, Mass.: Harvard University Press.

Wood, F. G., Jr. (1952). Porpoise sounds. Underwater sounds made by *Tursiops truncatus* and *Stenella plagiodon*. *Bulletin of Marine Sciences of the Gulf and Caribbean* 3: 120–133.

Wood, F. G., Jr. (1953). Underwater sound production and concurrent behavior of captive porpoises *Tursiops fruncatus* and *Stenella plagiodon*. *Bulletin of Marine Sciences of the Gulf and Caribbean* 3; 120–133.

Wu, Chau H. (1984). Electric fish and the discovery of animal electricity. *American Scientist* 72: 598–607.

Zakon, H. H. (1986). "The electroreceptive periphery." In T. H. Bullock and W. Heiligenberg (eds.), *Electroreception*, pp. 103–156. New York: John Wiley and Sons.

Index